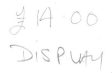

D0541089

LABORATORY ANIMAL HANDBOOKS NUMBER 14

The Design of Animal Experiments

Reducing the use of animals in research through better experimental design

Michael F W Festing
MRC Toxicology Unit, University of Leicester, PO Box 138, Lancaster Road, Leicester LE1 9HN, UK

Philip Overend
GlaxoSmithKline plc, New Frontiers Science Park (North), Third Avenue, Harlow, Essex CM19 5AW, UK

Rose Gaines Das
NIBSC, Blanche Lane, South Mimms, Potter's Bar, Hertfordshire EN6 3QG, UK

Mario Cortina Borja
Centre for Paediatric Epidemiology and Biostatistics, Institute of Child Health, University College London, 30 Guilford Street, London WC1N 1EH, UK

Manuel Berdoy
Oxford University Veterinary Services, Parks Road, Oxford OX1 3PT, UK

laboratory animals limited

The ROYAL SOCIETY of MEDICINE PRESS Limited

Published in the United Kingdom on behalf of Laboratory Animals Ltd (*www.lal.org.uk*) by The Royal Society of Medicine Press Limited, 1 Wimpole Street, London W1G OAE (*www.rsmpress.co.uk*).

British Library Cataloguing in Publication Data
A catalogue record for this book is available from the British Library
ISBN 1-85315-513-6
ISSN 0458-5933

Typeset and printed in Great Britain

Contents

Acknowledgements

This book was written as a collaboration between individuals representing their own organizations and/or working on behalf of the organizations listed below:

The Editorial Group was Chaired by **Bryan Waynforth**, GlaxoSmithKline plc but also representing the Institute of Biology, London, UK

The Secretary to the Editorial Group was **Ms Elizabeth Lester**, Institute of Biology, 20–22 Queensberry Place, London SW7 2DZ, UK

Michael F W Festing, The Fund for the Replacement of Animals in Medical Experiments (FRAME), Russell and Burch House, 96–98 North Sherwood Street, Nottingham NG1 4EE, UK

Philip Overend, The Institute of Biology, 20–22 Queensberry Place, London SW7 2DZ, UK

Rose Gaines Das, The British Region of the International Biometric Society

Mario Cortina Borja, The University of Oxford

Manuel Berdoy, The University of Oxford

Preface

It is universally accepted that persons who aspire to use animals for experimental purposes should receive proper training. At the heart of this training is the application of Russell and Burch's 3Rs, i.e. Replacement, Reduction and Refinement (Russell & Burch 1959). Of these, it is probable that least emphasis is placed on, and least success has been obtained in, reducing the number of animals that are currently used in animal experiments.

A major factor in bringing about a reduction in animal use is the correct application of experimental design and statistical analysis, both of which are, arguably, poorly taught or understood by undergraduates and postgraduates in the biomedical sciences. This deficiency is a major concern to a number of bodies, not least to the UK Home Office, which mandates courses (Modules 1–5) for the training of persons in preparation for applying for a licence to use animals; to the Institute of Biology (IOB), which is one of the two accrediting bodies for these courses; and to FRAME (Fund for the Replacement of Animals in Medical Experiments), which has set up a 'Reduction Committee' to recommend a resolution to the problem. Therefore, in order to redress this imbalance, the IOB decided to commission a book to provide, in a clear, concise and simple way, an understanding of the benefits to be derived by investigators in the proper design of their studies. It is expected that, in many cases, this will lead to a reduction in the number of animals needed, but in all cases it would lead to the optimum use of animals that would provide valid results.

It is not intended that this book should emulate the many textbooks that are available which give a detailed description of experimental design and statistical analysis. Rather, its purpose is to act as a teaching aid to impart an understanding of what types of experimental design and statistics should be considered when developing an experimental study protocol. References to textbooks and statistical packages are then given to enable these to be carried out. Undoubtedly, the best course of action when contemplating designing a study is to seek out and employ a statistician knowledgeable in the field of interest. He or she will want to know many of the details discussed in this book. It is hoped that the book will be of interest in all fields of scientific research and that it will achieve its aim of helping to improve studies such that animal use is reduced, whilst still ensuring that the maximum benefits are derived from the study.

Updates and corrections to this book, and additional Web sites of interest, will be posted on the Laboratory Animals Ltd Web site (*www.lal.org.uk/hbook14.htm*).

Bryan Waynforth
Chairman, Editorial Group

1

Basic principles of the design of animal experiments

Progress in medical and biological research is heavily dependent on the use of experimental animals. However, for both ethical and economic reasons it is vitally important that these animals are used in well designed experiments which give valid answers from the use of the minimum possible number of animals. An experiment that used twice as many animals as necessary to achieve its scientific objectives, or one so badly designed that it gave the wrong result, would be unethical if it involved pain or suffering of any animal.

This book is aimed at all research workers using experimental animals (primarily vertebrates), and may also be useful for people involved in the ethical review process. It is not a textbook on experimental design or statistical analysis, but it aims to cover the broad principles of design in a non-mathematical way, with specific reference to the ethical use of laboratory animals. 'Experimental design' is considered in a wider context here, as it includes consideration of the choice of animals and the control of variation—topics not normally covered in a statistics textbook. For those people who already have some knowledge of statistics and experimental design, it should help in the design of better, more efficient experiments that use fewer animals per unit of knowledge gained. For others, it should give them background information which will help them to consult a professional statistician/ biometrician more effectively.

There is indeed some cause for concern. Experimental design is a subject that all too often falls between two stools. Statisticians may know all about it in theory, but they seldom do experiments and may be uneasy teaching and promoting a subject in which they do not have first-hand experience. Scientists, on the other hand, do experiments all the time, but often have not been taught many of the basic statistical principles of design. All too often they simply repeat the type of experiment taught to them by their graduate supervisor, or which they see in the literature. They are often concerned that if they submit papers with unusual experimental designs (say with fewer than six rats per treatment group) these will be rejected by the referees, who themselves often have little training in design. Thus, a vicious circle develops which prevents progress and leads to a waste of animals.

Surveys of published papers suggest that there is ample scope for improvement. A survey of experiments published in two toxicological journals found that about a third of experiments used twice as many animals as would have been appropriate using the 'resource equation' method of determining sample size (Festing 1996). On average, the experiments used between 50% and 100% too many animals. The technique of blocking to increase the precision of an experiment is common in agricultural research, but in this sample of 48 experiments it was only used once, and then the authors did not carry out the correct statistical analysis. Factorial designs, in which more than one factor, such as dose level of a compound and time of sacrifice, are studied in a single experiment were quite common, but rarely did the authors use the correct statistical analysis. Overall, inappropriate statistical methods were used in more than 60% of papers. Other surveys provide similar findings (Altman 1991, McCance 1995, Porter 1999), though most concentrate on incorrect statistical analysis, rather than incorrect experimental design, and cover clinical rather than animal experiments.

Legal requirements and the '3Rs'

It is a legal requirement in the European Union that a scientist planning a research project that might involve vertebrates (and some other species such as *Octopus vulgaris*) must consider whether its objectives could be achieved using alternative methods. Although this legal requirement may not exist in other countries, it is ethically and economically desirable to consider alternatives. A useful approach is by way of Russell and Burch's '3Rs'—*Replacement, Reduction* and *Refinement* (Russell & Burch 1959)

The first step is to consider whether it is possible to use a 'Replacement' alternative. Could part or all of the research be done using lower organisms such as *Drosophila* or *C. elegans*, or using tissue culture or even mathematical modelling? If so, then those methods should be chosen in preference to using vertebrates.

If it really is impossible to use a replacement alternative, the next step is to consider 'Refinement'. The aim is to minimize pain, suffering or lasting harm to each individual animal. All animals, whether or not they are currently being used for an experiment, need to be comfortably housed with adequate space and a good diet. The Institute of Laboratory Animal Research of the National Research Council (NRC) in the USA and the Home Office in the UK provide guidelines or codes of practice (NRC 1996, Home Office

1995). The animals need to be cared for by trained staff with ready access to a veterinary surgeon. They should be free of clinical and subclinical disease. Animals should be handled regularly and sympathetically so that they do not feel fear when entered into an experiment. When used in an experiment every care should be taken to minimize pain and suffering. Where substances are administered to animals, procedures that cause the least possible pain should be used (Morton *et al.* 2000). Surgery should be done with appropriate anaesthesia and analgesia, with good postoperative care. Some experiments are expected to result in substantial pain or the death of some of the animals; in such cases humane endpoints should be used (Stokes 2000; *see also* other papers in the same issue of the *ILAR Journal*). Animals which develop tumours should be painlessly killed before the tumours become too large. Often it is possible to predict with some degree of confidence from an animal's appearance and behaviour that it is going to die; such animals must be painlessly killed rather than being left to die in pain. Finally, when the experiment is over the animals should be killed using an appropriate painless method such as one of the humane methods recommended by the UK Home Office (Home Office 1995, 2000) or the Institute of Laboratory Animal Research (NRC 1996).

Reduction, the main topic of this book, is concerned with minimizing the numbers used in each experiment, consistent with achieving the desired scientific objectives. Very briefly, it involves having a clear understanding of the objectives of the study, understanding and controlling variation with efficient experimental designs, extracting all the useful information in the experimental data by appropriate statistical analysis, and careful interpretation of the results. It is highly correlated with good scientific practice.

Overview of the steps involved in designing a good experiment

Scientists who do not have much training in experimental design and statistics should consider discussing their proposed experiment at an early stage with a statistician or colleague with a good background in statistics. The basic steps in the design process include the following:

Formulation of an hypothesis or other objective of the study

The aims of the experiment need to be clearly defined and stated explicitly. Many experiments are set up formally to test some relatively simple hypothesis and/or to estimate treatment means, differences between treatment means or the relationship between two variables such as dose of a chemical and an observed response. Exploratory experiments, on the other hand, may be aimed at gathering sometimes substantial amounts of data in order to formulate new hypotheses. Formal significance testing, though it is done, may not be the main objective of the study. Whatever the aims of the experiment, the design and subsequent statistical analysis are essentially the same process. No scientist should design an experiment without having a very good idea of how he or she intends to analyse the results.

Where animals or *in vitro* methods are being used to model humans, these will have been chosen because it is thought that they are able to tell us something about humans. In this case the hypothesis under test must be whether the chosen model responds to the treatment. It is not possible to use them to answer directly whether or not humans respond.

Choice of the 'experimental unit' and the need for independent replication

The 'experimental unit' is the unit of replication which can be assigned at random to a treatment. Different experimental units must, in principle, be capable of receiving different treatments. Very often the unit will be an individual animal. However, if the experiment involves comparing different diets, and animals within the same cage must have the same diet, then the experimental unit would be the cage of animals and the metric used in the statistical analysis would be cage means of the characters of interest. If lymphocytes were taken from an animal and placed in several culture dishes which were treated in different ways, then the experimental unit would be the individual dish of cells. In a crossover experiment (see Chapter 4) an animal may be assigned to a treatment for a period of time, then rested and assigned to another treatment. In this case, the experimental unit is the animal for a period of time. If an animal can be shaved and several different topical applications of a compound can be made to different patches of skin, then the experimental unit is the area of skin.

In 'split-plot' experiments there may be two different types of experimental units. A diet experiment may involve some cages with a control diet and some with an experimental diet. Within each cage there may be two rats. These might be injected either with saline or with a vitamin supplement. Thus, for comparing the diets the cage is the experimental unit, and for comparing the effects of the vitamin injection, and any interaction between diet and vitamin treatment, the individual rat is the experimental unit.

Failure to identify the experimental unit correctly is one of the most common mistakes in the design of animal experiments. Its effect is to render the resulting P values from the statistical analysis virtually meaningless, often making them unrealistically low. A well known example is the assumption that the individual pup in a teratogenesis experiment is the experimental unit, rather than the pregnant female, though clearly it was the female which was assigned to one of the treatments. Similarly, in studies of the effect of environmental enrichment in which cages of mice are assigned at random to different enrichment treatments, the statistical analysis should be done on the mean values for all animals in the cage, rather than on the values of individual animals within each cage.

Controlling variability

Controlling variation is vital. This is covered in Chapter 3. Generally, the more uniform the animals (or other experimental units) are, the fewer of them are needed to do an experiment. Uniformity can be achieved by choosing genetically uniform, isogenic animals where these are available, and by ensuring that they are free of clinical and subclinical disease, which inevitably increases variability. Animals housed under optimum environmental conditions tend to be more uniform than those housed in poor conditions. Mice housed singly appear to be more variable than those housed in groups (Chvedoff *et al.* 1980). When setting up an experiment the animals should be, as far as possible, of uniform weight and age. Where this is not possible, they could be stratified by weight, using a randomized block experimental design, or covariance analysis could be used to correct to a uniform weight (Chapter 4). Randomized block designs can also be used to take account of any natural structure in the available experimental material. For example, experiments with transgenic animals may be complicated by lack of availability of sufficient animals at the start of the experiment. However, provided enough animals are available to have one animal for each treatment group, the experiment could be done over a period of time as a randomized block design, where each block is a 'mini-experiment' with one observation in each treatment group.

Treatments should be applied uniformly to all animals. If measurement error is likely to be important, the use of multiple determinations should be considered. Any remaining variability should be distributed among groups using appropriate randomization (see Chapter 3). Stringent precautions need to be taken to avoid bias due to one group having a more favorable environment (in its broadest sense, i.e. including the environment when the measurements are made) than another. As far as possible, animals and samples should be coded so that staff are 'blind' to the treatment group.

Choice of treatments (independent variables)

The next step is to ensure that appropriate treatments (independent variables) are used. Usually one or more control groups which are untreated, sham-treated, or treated with a placebo should be included in the experiment. Controls are considered to be one of the treatment groups. If experiments are repeated over time, controls might be included at each time point in order to estimate time trends or fluctuations, though in these circumstances a randomized block design may be appropriate. However, historical controls from a previous experiment are usually of little value because there are many uncontrollable factors which can influence animals, such as seasonal influences, fluctuations in diet, microflora, personnel and methods of measurement, and there is no assurance that these will be the same in the groups being compared. Historical controls may, however, be useful if similar experiments are done repeatedly so that a substantial number of such experiments can be studied. In such circumstances industrial type control charts or 'Shewhart' charts which plot individual or mean results in successive experiments may be helpful (Dixon & Massey 1983).

Where a dose-response relationship is being studied, it is sometimes helpful in the subsequent analysis if the dose groups can be equally spaced on some scale. Linear and nonlinear trends can then be estimated within an analysis of variance (Altman 1991, Snedecor & Cochran 1980).

The number of treatment combinations also requires some thought. Mead (1988) suggests that most experiments should have between 10 and 50 different treatment groups. This is more than in most animal experiments, though it is not uncommon to have, say, four dose groups with three different time periods, giving a total of 12 treatment combinations. Mead suggests that anyone planning an experiment should get the maximum amount of information out of it, and in many cases this implies the need to find out not only whether the treatment groups differ (and by how much), but also what factors such as the sex or strain of animals influence these results. Factorial designs (Chapter 4) can be used to explore the interrelationships between independent variables such as treatment, sex and strain without having to use large numbers of animals.

Choice of dependent variables (characters to be measured)

Quantitative (measurement) data are usually more informative than categorical data. However, the main design principles remain the same whatever the nature of the data, and most of the numerical examples in this book assume quantitative data. Generally, it makes sense to measure as many things as is conveniently possible, though unless it is an exploratory experiment not involving formal statistical tests, the statistical analysis should concentrate on a few characters which are of particular interest. In some experiments, such as those involving the measurement of haematology and clinical biochemistry, there may be 10–20 or more dependent variables. In some toxicology experiments each rat produces a hundred or more observations, and with gene micro-arrays or electronic monitoring of brain function there may be thousands of observations on each experimental unit. Thus, at this stage the research worker needs to have a good idea of how the data are to be statistically analysed and interpreted once the experiment is completed.

Choice of a design

There are formal, properly randomized experimental designs that are suitable for every situation. The very simplest is the *completely randomized* design in which experimental units are allocated to a treatment at random. This design can have any number of treatment groups, it is relatively unaffected by unequal numbers in each group, and it is easy to design and analyse. However, it may be inefficient if the experimental material is heterogeneous or has some sort of natural structure. More advanced designs can be used to increase precision by using techniques such as blocking, or to increase the amount of information from a given sized experiment using factorial or split plot designs.

The *randomized block* design is used when there is some environmental factor which is uncontrollable (such as day-to-day variation), when the material is more heterogeneous than would be ideal, or when the experimental units have some natural structure. For example, if the experimental unit is a skin patch to test cutaneous inflammation on the back of a rabbit, where only a few patches are available on each rabbit, and several rabbits need to be used, then the rabbit represents a block of experimental units. A *Latin square* design is used when there are two sources of uncontrollable variation, such as day-to-day variation and minute-to-minute variation which both need to be taken into account.

Split-plot and *repeated measures* designs involve two different types of experimental units. The main comparison might be between animals which can be assigned to, say, three dietary or drug treatments designated A, B and C. However, each animal may also be assigned to a further set of treatments over a period of time. Treatment X, for example, could be given for a short period, followed by treatment Y. The order of X and Y could be assigned at random. Thus, with this design the whole animal is the main plot, and the animal for a period of time is the sub-plot. The terms are derived from agricultural field experiments. The term 'repeated measures' tends to be used when some character is observed repeatedly over a period of time, but the observations are expected to be approximately independent of each other. Further details are given in Chapter 4.

Other designs include *crossover*, *incomplete block* and *sequential* designs. Some designs are rarely used, but are important in special circumstances.

Factorial designs are ones in which the effects of two or more independent variables, such as a drug treatment and the sex of the animals, are studied in one experiment. Strictly, a factorial design is an arrangement of treatments rather than a 'design'. Thus, the factorial design is superimposed on the other designs mentioned above, so that it is possible to have a completely randomized factorial design, a randomized block factorial design, etc. All these designs are discussed in more detail in Chapter 4.

Determination of sample size

The experiment needs to be large enough to obtain a good estimate of the within-group standard deviation and of any treatment comparisons which are of interest, but an excessively large experiment will waste resources and animals. If there are many treatment groups, then the size of each group can be reduced. This is particularly true with factorial designs, where group sizes can often be quite small, yet the experiment may still be more powerful than a single factor one. The *power analysis* and the *resource equation* are the two main methods of determining sample size. These are considered in more detail in Chapter 5. Scientists need to make a real effort to move away from the defensive habit of using six animals per group just because everyone else does so. If there are more than five or six treatment groups, this often leads to a waste of scientific resources which could be better employed doing more experiments and thereby speeding research progress.

Statistical analysis planning

Scientists should have a good idea of how they plan to do the statistical analysis *before* starting the experiment, and each experiment should be analysed before starting the next one. Failure to do so may mean that a whole series of experiments are done incorrectly or in such a way that the results are not analysable.

Some thought needs to be given to any possible complications from missing data due to unexpected loss of animals. Factorial designs in which a whole group is missing can cause problems so, if many missing data are expected, the design may need to be modified.

Pilot study, if necessary

If severe technical problems are encountered in a large experiment, the animals and scientific resources may be wasted. Thus, when the experiment involves difficult techniques that are new to the scientist, or complicated logistics, then a pilot experiment can be used to ensure that it is technically feasible. These pilot experiments may provide useful information for planning a more definitive study, and can involve small numbers of animals. However, small experiments provide imprecise estimates of within-group variation, so care must be exercised in using such data in a power analysis to determine sample size in the main experiment.

Protocols and standard operating procedures (SOPs)

Once the experiment has been designed, protocols need to be prepared and discussed with all other staff involved. In particular, the experiment needs to be explained in detail to the animal house staff. They will be more cooperative and helpful if they understand the purpose of the experiment.

A discussion of ways in which the experiment is implemented is beyond the scope of this book. However, a few points need to be mentioned. A diary should be kept with notes of any unexpected events. Loss of animals and the reasons for the loss should always be recorded and noted in the published report. If there is reason to be suspicious of a particular observation or measurement, this should also be noted. If, in the final statistical analysis, the data point is found to be a serious outlier then it is useful to know that it was judged to be unreliable at the time the observation was made. However, an observation should never be discarded just because it is an outlier. One procedure is to analyse the data with and without the observation(s) in question, with an appropriate transformation of the data if necessary. If the conclusions are the same, then the data can be retained. If the conclusions are different, and there are no firm grounds for discarding the observations, it is probably unsafe to draw any very firm conclusions, and it may be necessary to repeat the experiment. Non-parametric and so-called 'robust' statistical methods may also be considered when the material is seriously heterogeneous.

Statistical analysis

Finally, the data will need to be analysed and interpreted. Many experimental designs assume that the experiment will produce quantitative data that can be analysed by the analysis of variance (ANOVA), possibly after a transformation of the data to an appropriate scale with equal variances within each group and with normally distributed residuals (these are the differences between group means and the individual observations). Normality and homogeneity of variances should normally be examined as part of the statistical analysis. Scientists planning to do research with laboratory animals should be familiar with the ANOVA (Chapter 4), which is a powerful and general method of analysing quantitative data.

Interpretation

The aim of statistical analysis is to give some quantitative estimate of the probability that any observed differences between treatment groups could have arisen by chance sampling variation due to differences among individuals, rather than being caused by the treatments. Where it is unlikely that a difference could have arisen by chance, it is possible to obtain some idea of the likely magnitude of any differences by estimating a confidence interval for the difference. However, the statistical analysis will not normally give any indication of the generality of the result, i.e. the extent to which the same result could have been obtained under different conditions. Where characteristics such as the strain and/or sex of the animals have been included as factors in a 'factorial' experimental design, the experiment may show whether the result is strain- or sex-dependent. Similarly, the use of a randomized block design, which breaks the experiment up into a series of 'mini-experiments', will also help to increase the generality of the results because treatment differences will only show up as being statistically significant if they are found reasonably repeatably in each block. Nevertheless, the interpretation of the results is largely a matter of scientific judgement, though the statistical analysis should prevent claims of important treatment effects when these could be due entirely to sampling variation.

Presentation of the results

The results should be presented as clearly as possible. The experimental design used (e.g. 'completely randomized', 'randomized block', 'split-plot', etc) should be described in the materials and methods section of the paper. The method of statistical analysis should be stated, with appropriate references if the methods are in any way unusual. Tables and figures should be clearly laid out. If the variances are not different in each group (as shown by a study of the 'residuals', Chapter 4), then it will usually be appropriate to give a pooled estimate of the standard deviation, rather than calculating a standard deviation individually for each group. As far as possible sufficient numerical data should be given to allow other investigators to reach the same conclusions as the authors.

Sometimes results are not statistically significant, but this is, in itself, of interest. For example, an experiment may show no evidence that a particular compound is toxic at the levels given. However, editors may be reluctant to publish papers showing non-significance, even if it is important. Non-significance may occur because there is no true effect, or because the experiment was too small to be able to detect a biologically important difference. A power analysis (Chapter 5) can often be used to determine what power the actual experiment had to detect a specified effect. If the experiment was actually a powerful one for the specified effect size, this can be stated and the paper may therefore become acceptable for publication.

Common errors in the design and statistical analysis of experiments

There have been several studies of statistical errors in published papers, though most have been concerned with clinical research rather than animal studies (Altman 1982, 1991; Festing 1994a, b). However, few of these have investigated errors in experimental design. One problem is that authors seldom give sufficient information on exactly how the experiment was done for it to be possible to judge whether it was done well. For example, authors rarely state whether the experimental material was assigned to the treatment groups at random. Presumably referees either assume that this has been done, or they consider it relatively unimportant, so they accept the paper. The common errors listed below are based on the experiences of the authors of this book, rather than on a formal analysis of published studies.

Errors in design

Experiments done on an *ad hoc* basis

Sometimes experiments are not designed at all. Treated and control groups may well be separated in time or space with new groups being added on the whim of the investigator, so that any comparison could be confounded by other variables. One clue that this might have happened is when the group sizes vary for no apparent reason. Why did the control group have five, but the treated group have 17 rats? Well designed and correctly randomized experiments should give reasons for the group sizes used. The standard error of the difference between two means is lowest if the two groups are of equal size for any given number of subjects. However, sometimes it is better to have more animals in the control group if it is being used in several comparisons. It is not uncommon to see tables with a footnote saying something like '$n = 3$–6' without any further explanation. It is essential for all experiments to be designed before they are started.

Inappropriate choice of treatment groups

Factorial designs are sometimes unbalanced with one treatment combination missing, so that the correct factorial statistical analysis is difficult or impossible. Controls may not exactly match the treated group, and in some cases both vehicle and untreated controls are included, when only the former is strictly relevant. Mead (1988) suggests that as much information should be obtained from each experiment as is reasonably practical. He suggests that a well designed experiment will have between 10 and 50 different treatment groups. This is not difficult to achieve with factorial designs (see Chapter 4) by splitting treatment groups into subgroups, without any substantial increase in numbers of animals. However, very few animal experiments have so many treatment groups.

Experiment too large or too small

Very small experiments lack statistical power to detect biologically important results. However, increasing the size of an experiment beyond a certain point leads to diminishing returns. Thus, there is an optimum size for an experiment which is a balance between power and cost, where cost includes non-financial costs such as the suffering of the animals.

Failure to use blocking as a means of controlling variation

With many experiments it is patently obvious that there are bottlenecks such that the experiment needs to be broken up into smaller parts. It is not possible to bleed 24 rats from four different treatment groups all at the same time at five-minute intervals for a total of 10 bleedings, as shown in one paper. All too often experiments of this sort are divided up by treatment group, leading to possible biases. A much better alternative would be to use a randomized block design, which is rather like a series of mini-experiments each of which may only have a single experimental unit from each treatment group, as discussed in Chapter 4. In this example, only four rats, one from each group, would be handled at a time.

Continued use of genetically heterogeneous animals when isogenic strains are available

Many research workers continue to use genetically heterogeneous (outbred) animals in the mistaken belief that they represent a better model of genetically heterogeneous humans. Thus over 70% of 'rat' papers published in one toxicology journal in 1999 involved outbred rats. Unfortunately, genetic heterogeneity leads to phenotypic heterogeneity so that the power of the experiment to detect a treatment effect will be reduced. In effect, all that phenotypic heterogeneity does is to obscure the more subtle effects of a treatment. Many people who use these animals know that uniformity of the experimental material is important, and specify animals with a narrow weight range. For some inexplicable reason phenotypic heterogeneity caused by genetic variation is assumed to be different from phenotypic heterogeneity caused by environmental factors, and the former is considered to be 'good' while the latter is considered to be 'bad'. This is scientific nonsense. The use of isogenic strains as a powerful research tool is now well established, and the quicker all research workers move to the use of such strains, when available, the better. In fact, the use of genetically heterogeneous animals may seriously damage the quality of the research (Festing 1999). Funding authorities should note this.

Errors in the statistical analysis of experiments

There have been several surveys of the statistical quality of published papers (cited above). Most concluded that more than half the published papers had statistical errors of some sort, though these do not always invalidate the conclusions. Some of the most common errors are:

Failure to do any statistical analysis of numerical data

Papers continue to present numerical data without any statistical analysis, sometimes with the claim that there is insufficient data to be worth analysing. This may be true, but in many cases a statistical analysis is particularly important when there are few data in order to prevent unsupported conclusions from being reached. Percentage data are particularly prone to abuse because if the numbers in each group are small, percentages can differ widely but still not be significantly different. It is quite rare for percentages to be presented with their standard errors, or better still with a 95% confidence interval.

Failure to screen raw data for errors

Sometimes data contain obvious errors due to misreading of instruments or incorrect transcription. It is not clear how often such data remain uncorrected because raw data are seldom published. Errors of this type can often be detected using graphs of various types. As an example, data have been published on the Web giving the sizes of over 1000 microsatellite markers in two strains of mice which should be genetically almost identical. When graphing size in one strain versus the other, most of the observed differences were small and clearly due to measurement error. However, 28 of the differences were large, and some almost certainly transcription errors, such as the size being given as 198 base pairs in one strain and 98 base pairs in the other (Festing 2000).

Inappropriate use of Student's *t-test*

Student's *t-test* is a parametric test which is suitable for comparing the means of two groups with approximately equal variances and with the residuals (i.e. the deviations of each observation from its group mean) having a normal distribution. Although there are versions of it which can be used with unequal variances and non-normality, it is generally better to use a non-parametric test such as the rank-sum (Mann-Whitney) test, or to transform the data to a scale where normality and equal variances apply (e.g. by taking the log or square root of the observations, depending on circumstances).

Student's *t-test* is often incorrectly used to try to analyse experiments with more than two treatment groups, and it is quite inappropriate for analysing factorial designs. The problem with using the test in this way is that with, say, four groups there are six possible comparisons that can be made between means. This implies six tests, each of which has a chance of giving a false-positive result. If a 5% significance level is chosen, there is approximately a 26% chance that one or more of the results will be 'significant' even if the treatment is having no real effect. It is possible to use a Bonferroni correction, but that tends to be too conservative, giving too many false-negative results.

A second problem is that the *t-test* lacks statistical power because each test is conducted on only a subset of the data, i.e. the two groups actually being compared. This ignores all the useful information in the other groups on within-group variation. This means that the *t-test* will often fail to pick up biologically important effects which would have been 'significant' if a more appropriate test had been used.

Finally, the *t-test* is not suitable for analysing factorial designs because it is not easy to use it to test for interactions. All these problems can usually be overcome using an analysis of variance, described in Chapter 4.

Misinterpretation of *P* values

Lack of significance is sometimes equated with there being no treatment effect. However, 'absence of evidence is not evidence of absence'. A non-significant difference may mean that the treatment has no effect, or it may arise from the fact that the experiment is incapable of detecting an effect which might still be sufficiently large to be of biological significance. This can happen if the material is highly heterogeneous, causing noise which obscures the treatment effect, or if the experiment is too small. The exact meaning of *P* values is also commonly misunderstood. A *P* value is not the probability that the null hypothesis is true. It is the probability that a result at least as extreme (say as a difference between means) as the one actually found could have been found if the null hypothesis were true.

Inappropriate or incorrect statistical analysis

Sometimes the statistical analysis is simply not appropriate. In one study the liver weight of partially hepatectomized rats was compared with controls, using a *t-test*, immediately after half the liver had been removed. It is trivial to test the hypothesis that half a liver weighs less than a whole liver. In other cases the wrong test is used, or the calculations are incorrect. For example, an attempt is made to do a chi-squared test on percentages, rather than counts. The chi-squared test is sometimes used when the counts are too low. It is only accurate if the expected values in each cell are five or more. If this is not the case, then Fisher's exact test should be used.

Unexplained omission of data

Observations are sometimes discarded without explanation because they spoil what would otherwise be good results. Where sample sizes have not been justified, and are unequal for no obvious reason, then there might be a suspicion that individual data points have been discarded. Where outliers occur, their cause should be studied as far as possible. In some cases they are simple recording errors, in which case it may be possible to make corrections; if not, then the fact that some data have been omitted and the reason for this should be noted. In some cases outliers and unusual observations can be highly informative, so they should not simply be discarded.

2

Choice of animals

Introduction

Whenever possible, animal experiments should be done using high quality animals which are free of clinical or subclinical disease, 'genetically defined' (i.e. isogenic, mutant or transgenic), and maintained on a nutritionally adequate diet and in a good environment. Such animals will be more expensive than 'conventional' animals (a term used by laboratory animal scientists to describe animals of undefined health status), but fewer will be needed and the results will be more repeatable and reliable. Given that the cost of the animals usually represents only a small fraction of the total cost of the research, it makes no ethical or economic sense to buy cheap animals if that reduces the scientific quality of the work.

Freedom from disease

Before about 1950 all species of laboratory animals carried a wide range of viral, bacterial and metazoan organisms. These caused clinical or subclinical disease which disrupted research by decreasing lifespan, increasing variability so that more animals were needed, and in some cases by interacting with the experimental treatments to give spurious results. Thus, the average lifespan of a laboratory rat in the 1930s was only about 12 months compared with 2–3 years, depending on strain, today. This was largely due to infections with *Mycoplasma pulmonis* and other microorganisms that caused chronic respiratory disease. At one stage it was thought that vitamin A deficiency caused lung damage because rats which were vitamin A deficient had more serious lung lesions than non-deficient rats. However, it was subsequently found that the physiological stress of vitamin A deficiency was increasing the frequency of lung lesions normally associated with chronic respiratory disease (Lindsey *et al.* 1971).

Because animals with clinical or subclinical disease are usually more variable than healthy ones, more are needed in an experiment to achieve a given level of statistical precision. In one study (Gartner 1990), the standard deviation of kidney weight of rats suffering from chronic respiratory disease was 43.3 (arbitrary) units, whereas it was only 18.6 units in healthy rats. A *power analysis* (described in more detail in Chapter 5) can be used to decide how many animals would be needed if an experiment were to be conducted to study the effect of some treatment on kidney weight. Assume that a 10% difference between the treated and control groups would be of interest, and the experiment is to have an 80% power (i.e. an 80% probability of detecting such an effect) with a 5% significance level. It is estimated that the experiment would need 42 rats per group using the healthy rats or 200 animals per group (five times as many) using the ones with chronic respiratory disease. Moreover, the conventional rats would suffer both from the experimental treatment and the respiratory disease. It is simply not ethical to use such animals.

In the late 1950s so-called 'Specific Pathogen Free' or 'SPF' techniques were developed as a means of eliminating pathogenic microorganisms. Fetuses *in utero* are usually microbiologically sterile. These were removed from their mother just prior to parturition, under sterile conditions, and hand reared in an isolator using sterile milk and diet. This resulted in the establishment of germ-free or 'gnotobiotic' colonies of mice and rats. Although these animals survived and bred, they were abnormal in many respects, and they were difficult and expensive to maintain. Further research showed that animals infected with a suitable cocktail of non-pathogenic gut bacteria became like normal conventional animals which were free of disease. Colonies of these animals were later established in 'barrier' animal houses in which all supplies of diet, bedding and equipment were sterilized to prevent re-infection with pathogens. Although these SPF animals sometimes pick up unwanted microorganisms, often from the staff, they are essentially like conventional animals which are free of all important pathogenic microorganisms. It is both economically and ethically desirable that these animals should be used wherever possible as, although they cost more than conventional animals, far fewer are needed, and there is less likelihood that the treatment effects will be confounded with the effects of pathogenic microorganisms.

SPF mice, rats, rabbits, and some other species are now widely available from commercial breeders. These should have been regularly screened and found to be free of certain defined viruses, bacteria and parasites, details of which will be given by the breeder on request. The pharmaceutical industry has recognized the importance of using such animals and has invested heavily in good quality animal housing, but some academic establishments continue to maintain conventional animals. These are a source of infection for any SPF animals which are brought in to the animal house. There is a particular problem with transgenic strains produced in an academic animal house that is not of SPF standard. There is already evidence that they are spreading disease when they are distributed to other investigators around the world. This problem may also limit collaboration between institutes.

Choice of species

Many studies use a particular model of a disease such as cancer or atherosclerosis, or are designed to investigate the effect of knocking out a particular gene. Choice of species will then be dictated by the availability of the model in each species. Where there is a choice, the usual strategy is to use the mouse: it is small and economical and there are many strains, mutants and techniques available for the species. The rat is another obvious choice: it is larger, and there is extensive experience of its use in physiology, pharmacology, psychology and toxicology. Other rodents such as the hamster, guineapig or gerbil, or a larger species such as the rabbit, dog or pig are also widely used and are well adapted to laboratory conditions. The choice of species has some implications for the design of individual experiments. Isogenic strains (see below) of mice and rats are readily available, and their phenotypic uniformity and other useful characteristics mean that experiments can be designed using fewer animals than if non-isogenic animals are used.

It is noticeable that the larger the animal, the fewer tend to be used in each experiment. In an unpublished survey of papers in two toxicology journals, 11 studies using mice averaged 227 mice per paper (median 110, minimum 51, maximum 839), whereas the 44 studies using the rat averaged 74 rats per paper (median 39, minimum 4, maximum 931). Generally, there is no obvious statistical justification for using fewer rats than mice in similar experiments with similar objectives. However, if more characters can be measured, or if they can be measured with greater precision using larger animals, if lower power experiments are more acceptable, or if only large treatment effects are of interest, then it may be possible to justify the use of fewer, larger animals. The exact reasons for the differences need further investigation. However, because of their small size and wealth of genetic variants, mice are often used in preference to rats for genetic studies. Genetic mapping, mutagenesis and the genetic analysis of characters with a complex mode of inheritance all require larger numbers of animals than the formal controlled experiments in which the rat is so commonly used.

The *UFAW Handbook on The Care and Management of Laboratory Animals* (Poole & English 1999) is a useful source of information on the characteristics of individual species, and the *Principles of Laboratory Animal Science* (van Zutphen *et al.* 2001) gives more general information on laboratory animal science without having chapters on individual species. A brief description of some species characteristics is given in Table 2.1a & b.

Table 2.1a   Some species characteristics

Character	Mouse	Syrian hamster	Rat	Guineapig	Rabbit
Typical adult weight (g)	25–30	30–40	250–600	500–800	1000–7000
Average longevity (years)	1–3	3	2–3	6–8	5–6
Chromosome number	40	22	42	64	44
Age at puberty (days)	35	45–60	45–75	45–75	150–210
Usual breeding age (days)	50	80	80	80	150–210
Usual breeding method[1]	PM	HM	PM/H	PM	HM
Length of oestrus cycle (days)	4–5	4	4–5	14–16	Induced ovulation
Gestation period (days)	20	21	21–23	65–72	31–32
Average litter size	6–9	4–6	6–10	3–4	6–8
Typical weight at birth (g)	1–2	1–2	5–6	85–90	100
Typical weaning weight (g)	10–12	6–8	35–40	250	1000
Age at weaning (days)	19–21	21	21	14–21	50

[1]PM = Permanently mated, HM = Hand mated, H = Harem mated

Table 2.1b   Some species characteristics

Character	Ferret	Cat	Dog	Mini-pig	Rhesus monkey	Marmoset *Callithrix jacchus*
Typical adult weight (kg)	0.75–0.80	3.5–4.5	10–30	20	60–70	0.24–0.60
Average longevity (years)	5	13–17	13–17	>6	15–20	8–12
Chromosome number	40	38	78	38	42	44–48
Age at puberty (days)	180–270	180–240	180–350	60–70	800	390–450
Usual breeding age (days)	180–270	270	270	80–100	1200	400
Usual breeding method[1]	H	HM/H	HM	H	H/HM	PM
Length of oestrus cycle (days)	–	15–28	22	21	28	28
Gestation period (days)	42	63	63	114	146–186	130–144
Average litter size	6–7	3–4	3–8	5–7	1	2
Typical weight at birth (g)	10	110	200–500	400–600	380–450	25–35
Typical weaning weight (kg)	0.45	0.7–0.8	1–2	4.5–6.0	1.0–2.5	50–100
Age at weaning (days)	50	50	42–50	56	120–180	80–140

[1]PM = Permanently mated, HM = Hand mated, H = Harem mated

Genetic definition

Research workers using mice and rats are faced with a bewildering range of 'genetic' types. These fall into three main classes:

(1)   Genetically undefined 'outbred' stocks.
(2)   Genetically defined 'isogenic' strains (including congenic, consomic and recombinant inbred and congenic strains and F1 hybrids).
(3)   Partially genetically defined strains (mutants and transgenes on an undefined genetic background).

Within each type, there are many individual strains and stocks each with its own specific characteristics. Thus, making a rational choice of a strain for a particular project is apparently a daunting task. However, in some cases a particular strain, mutant, or transgenic strain will be chosen because it has useful characteristics such as hypertension, a particular type of tumour, or a disease such as diabetes. Where a 'general purpose' animal is wanted, then consideration should be given to the use of one of the more common isogenic strains. The properties of the main classes of stock are discussed briefly below.

Genetically undefined outbred stocks

All individuals of this class of stock are genetically unique, so that little or nothing is known about the genotype of an individual unless it is specifically typed. Outbred rats and mice continue to be commercially available, and are still widely used in some disciplines, although a scientific case attempting to justify their continued use has rarely been published in the last 20 years. As research material they suffer from a number of serious defects.

(1)   They tend to be phenotypically more variable than isogenic strains, so more are needed to achieve a given level of statistical precision.
(2)   They are subject to serious genetic drift over periods of a few generations if not maintained to rigorous standards (Papaioannou & Festing 1980). This means that background data on strain characteristics may rapidly become obsolete.
(3)   Stock names such as 'Wistar' rats or 'Swiss' mice are virtually meaningless because Wistar rats from different sources can differ widely in their characteristics. Thus if research is not repeatable in another laboratory, it is never clear whether this is due to differences between nominally identical stocks or some other cause.
(4)   Little or nothing is known about the level of genetic variation, or the actual alleles present in individual animals. Some outbred stocks are genetically heterogeneous; others are nearly inbred as a result of restricted population size and the gradual accumulation of inbreeding over many generations.

(5)   There are no genetic tests that can be used to identify a particular stock. It is impossible to distinguish between, for example, Wistar and Sprague-Dawley rats.

Attempts are sometimes made to justify the use of outbred stocks on two grounds, neither of which stands up to critical examination. The first is that 'Humans are genetically heterogeneous, and the aim is to model humans, so it makes sense to use genetically heterogeneous animals'. However, this is a misunderstanding of the whole concept of models. Models do not have to resemble the thing being modelled in every respect. It is only important that the relevant aspects of the model resemble the thing being modelled. Thus, for genetic studies it is useful if mice and rats are good genetic models of humans. But for physiological, biochemical, or toxicological studies they should be good physiological, etc. models, but do not necessarily need to be good genetic models. Isogenic strains have many useful characteristics which make them a much more valuable research tool than outbred stocks, and for most studies there is absolutely no reason to suppose that they are worse models of humans just because the individuals within a strain are genetically identical.

Second, toxicologists and some other scientists sometimes suggest that a high level of genetic heterogeneity will ensure that there is a greater chance that at least some susceptible animals will be present in the test population used to screen for toxicity. However, although the phenotypic variation within an outbred stock is usually greater than that in an isogenic strain, differences between isogenic strains are substantially greater than the variation within a single outbred stock. Thus, the best strategy to get a wide range of genetic heterogeneity, while still designing a powerful experiment, would be to use small numbers of several inbred strains. In this way, treated and control groups will be genetically identical at the start of the experiment, giving high statistical power, yet the differences between strains would provide much better coverage of possible variation in susceptibility than could ever be obtained through the use of a single outbred stock (Festing 1995, 1997). An experimental design involving both a treatment, such as a potentially toxic agent, and more than one strain of animals is known as a factorial design. Such designs can be highly efficient (see Chapter 4).

Occasionally, people also claim that isogenic humans are not used in clinical trials, so isogenic animals should not be used. However, identical twins are isogenic, and if they were readily available, they would be ideal material for clinical trials. Identical twin cattle are sometimes used effectively in agricultural experiments in spite of the difficulty in obtaining them, and one of the benefits of being able to clone larger animals will be that they will be useful for designing more powerful experiments, requiring fewer animals.

Isogenic strains (inbred strains, F1 hybrids, congenic strains, consomic strains, monozygous twins)

The most important feature of these strains is that the genotype of each individual is replicated in one (in the case of monozygous twins) or more individuals. Inbred strains are produced by 20 or more generations of brother × sister mating, and are in many ways like immortal clones of genetically identical individuals. The strains tend to be phenotypically uniform (when compared with outbred stocks), they stay genetically constant for many generations, and most have been genotyped at many loci. There is substantial background information on the origin, history, genotype and phenotypic characteristics of each strain. They can easily be genetically authenticated from a small sample of DNA.

Over 400 inbred strains of mice and 200 strains of rats are available throughout the world, and many are commercially available. They represent the nearest thing to a pure analytical grade reagent that is possible with animals, and more than a dozen Nobel prizes have been awarded for work which depended on their use (Festing & Fisher 2000). About 80% of research using isogenic strains is done using the 10 most popular strains (Table 2.2). Each strain has its own unique characteristics, and strain differences can be found for almost any character that has been studied. Some strains have a high incidence of cancer, others of heart disease,

Table 2.2   Top ten inbred mouse and rat strains (estimated from number of holders, summing across substrains and ignoring congenic strains)

Rank	Mouse		Rat	
	Strain	Holders	Strain	Holders
1	BALB/c	162	F344	51
2	C3H	145	LEW	37
3	C57BL/6	116	SHR	34
4	CBA	113	WKY	31
5	DBA/2	92	DA	28
6	C57BL/10	60	BN	26
7	AKR	52	WAG	21
8	A	51	PVG	17
9	129	50	BUF	17
10	SJL	40	WF	15

others have neither. Some are active, others are passive. Some like and some detest alcohol. Some learn well and others not so well in a particular learning task. Naturally, care has to be taken to choose a strain which is appropriate for the particular research project. It would not be sensible to use the AKR mouse strain in a long-term carcinogen screening study because most of these mice will have died of leukaemia before they are a year old. For general research where there is no requirement for animals of a specific genotype, the best strategy would be to use one or more of the most popular strains such as BALB/c or C57BL/6 mice or F344 or LEW strain rats shown in Table 2.2. Where a series of experiments is planned it might be useful to do a small pilot study involving several strains to find one which responds most appropriately. It would be unwise to do a long series of experiments using a single strain without occasionally using a different strain to make sure that the observed results are not unique to the chosen strain. Lists of inbred strains of mice and rats with some of their phenotypic characteristics are available on the Web (www.informatics.jax.org), and the genealogies of inbred mouse strains have recently been updated (Beck *et al.* 2000).

F1 hybrids, the first generation cross between two inbred strains, are also isogenic, and have the advantage that they tend to be very robust and vigorous due to hybrid vigour. They can be used instead of, or in association with inbred strains. However, unlike inbred strains they will not breed true as they are heterozygous at all the loci at which the parental strains differed, leading to genetic segregation in later generations. This can cause problems because, due to their vigour and good breeding performance, F1 hybrids are often used in the production of transgenic strains. Unfortunately, the resulting transgenic stock will be segregating at many loci, which can cause variation in expression of the transgene and serious genetic drift in the first few generations. Ideally, the transgene should be back-crossed for several generations to an inbred strain (Silva *et al.* 1997).

Anyone using isogenic strains is strongly urged to use the correct nomenclature so that it is clearly established which strains were used, and the work can be repeated in other laboratories. There are formal nomenclature rules (www.informatics.jax.org), but briefly strains are known by a code which consists of a few upper-case letters and sometimes a number (examples are shown in Table 2.2). While the rules are relatively well observed with mice, those working with rats seem to have an uncontrollable urge to find out the origin of the code and spell it out in full. This is in breach of the formal nomenclature rules, and should not be done.

Isogenic strains are usually phenotypically more uniform than are outbred stocks (discussed below). One result is that in most cases fewer animals are needed to achieve a given level of statistical precision. This is shown in Table 2.3 using data on hexobarbital sleeping time in mice (Jay 1955). It is estimated that an experiment set up to detect a 10% change in sleeping time between a treated and control group, under the conditions described in the table, would require between seven and 105 mice per group using one of the inbred strains, or between 144 and 257 of the outbred mice. Alternatively, if it was decided that only 20 mice could be used per group, then the power of the experiment to detect a 10% change in sleeping time would range from 28% to >99% using the inbred strains and from 14% to 23% using the outbred mice. These differences are, perhaps, more dramatic than usual, but they do serve to emphasize the point that usually fewer inbred than outbred animals are needed. The importance of understanding and controlling variation is discussed in detail in Chapter 3, and the power analysis method of determining how many animals are needed in each group is discussed in Chapter 5, using the data in Table 2.3 as an example.

Congenic, consomic and recombinant inbred strains are specialized types of isogenic strains developed largely for genetic studies. A pair of congenic strains are produced by backcrossing a defined genetic locus (the differential locus) from a donor strain to an inbred strain (the inbred partner). After about 12 generations of backcrossing the congenic strain will be genetically very similar to its inbred partner, but will differ at the differential locus. Any differences between the congenic strain and its inbred partner can

Table 2.3 Hexobarbital sleeping time in mice of five inbred strains and two outbred stocks. The table shows the strain name, class of stock, numbers tested, mean sleeping time, standard deviation (SD) and estimates of the number of mice that would be needed to detect a 10% change in the mean sleeping time in an experiment involving a control and a treated group of mice with sleeping time as the dependent variable, assuming the data are analysed using a two-sample *t-test* (2-sided) with a 5% significance level and a power of 90%. The table also shows the power of an experiment to detect such an effect if sample size is fixed at 20 mice per group. See Chapter 5 for more details of sample size using power analysis

Strain	Class	No. tested	Mean	SD	No. needed	Power[1]
A/N	Inbred	25	48	4	16	95
BALB/c	Inbred	63	41	2	7	>99
C57BL/HeN	Inbred	29	33	3	19	92
C3HB/He	Inbred	30	22	3	41	61
SWR/HeN	Inbred	38	18	4	105	28
CFW	Outbred	47	48	12	133	23
Swiss	Outbred	47	43	15	257	14

[1]to detect a 10% change in the mean assuming a fixed sample size of 20 mice per group

(with some reservations) therefore be attributed to the effects of the differential locus.

Congenic strains have been widely used in transplantation immunology (Snell & Stimpfling 1966), and many laboratories studying quantitative trait loci are now using these methods to isolate and study quantitative trait loci (QTLs), which control many characters such as cancer, hypertension and diabetes. Genetic markers can be used to speed up the backcrossing programme (Markel *et al.* 1997).

Sets of *consomic strains* are currently being developed in both mice and rats. These strains differ from an inbred partner strain for a whole chromosome, which is derived from a donor strain. They will also be used in the genetic dissection of characters controlled by QTLs (Nadeau *et al.* 2000).

Several sets of 'recombinant inbred strains' have also been developed from a cross between two standard inbred strains followed by 20 or more generations of brother × sister mating to produce a large number (preferably 20–30) of new recombinant strains. These can be used to map genetic loci which differ between the two parental strains (Taylor 1996). Further details of these and other genetic types are given by Silver (1995).

Partially defined strains

These strains are genetically partially defined in the sense that genotype at a particular locus is known, but the genetic background is undefined. Transgenic strains produced by incorporating foreign DNA into the genome of an animal (Maclean 1994), and induced or spontaneous mutations are often developed on a genetically heterogeneous genetic background because such mice and rats are more vigorous and breed better than isogenic ones.

Unfortunately, this can cause serious problems because over a period of a few generations there may be serious genetic drift in the background genotype which substantially modifies the expression of the transgene or mutant. Where the transgene is on a heterogeneous background, it is highly desirable to do a few generations of backcrossing to an inbred strain (as described above for congenic strains) before starting any detailed studies (Silva *et al.* 1997). Genetic drift over the first few generations may mean that scientific findings can no longer be repeated.

The importance of diet and environment

Bedding, diet and the physical environment can influence the outcome of an experiment. In the past, sawdust has sometimes been contaminated with insecticides used to treat timber, though reputable suppliers of bedding now generate their own from chosen timber known not to have been treated in this way. Paper from the office shredder should not be used as nesting material for mice and rats as it may contain inks that alter the response of the animals. Softwood sawdust contains substances that induce drug metabolizing enzymes, which can alter the response of mice to toxic agents. Thus, in one study it halved the sleeping time of mice under barbiturate anaesthetic compared with mice maintained on hardwood bedding (Vesell 1968), and in another it completely altered the response to a toxic chemical (Malkinson 1979).

The frequency with which animals are cleaned out may influence their responses. Mice and rats are uncomfortable if their cage is changed too frequently, but ammonia levels can build up to levels that might predispose the animals to respiratory disease if the cages are not cleaned out sufficiently frequently (Höglund & Renström 2000).

Diet is another possible source of contamination. If a large proportion of the farm animal feedstuff in Belgium can become contaminated with dioxin, as happened recently, something similar could potentially happen with laboratory animal diet. However, reputable laboratory animal diet manufacturers test the raw ingredients and the finished diet for a range of contaminants. Such diet is obviously more expensive, but a reliable and consistently uniform diet will result in more uniform animals and more repeatable experiments. This is certainly well worth some extra expense.

The diet can have an important influence on the characteristics of the animals, particularly those on long-term studies. Modern rodent diets should be nutritionally complete, though in some cases they have been formulated to maintain pregnancy and lactation, and may be too nutritionally dense for the long-term maintenance of non-breeding rodents. One result is that these animals may become obese and develop a high incidence of cancer and circulatory disorders, with a shortened lifespan. Unfortunately, it has proved to be difficult to formulate diets that prevent this obesity. There is now a move towards dietary restriction as a means of prolonging the life of animals on long-term studies (Masoro 1993).

The physical environment can also influence the animals. Mice housed singly tend to be more variable in body weight, and possibly other characteristics, than if they are housed in groups of 2–8 (Chvedoff *et al.* 1980). This means that more of them are needed in each experiment. Caging density may also influence the incidence and type of disease in mice (Les 1972). On the other hand, male mice housed in groups may start to fight, and in some cases they may cause each other serious damage. An experimental protocol should never involve regrouping adult male mice as they are almost certain to fight. Environmental enrichment, i.e. providing more complex cage environments, is currently being studied in a number of laboratories. It is too early to tell what effect this has on the animals, but some research suggests that this increases aggression among grouped male mice, which may increase variability (Nevison *et al.* 1999). The situation is complicated by strain differences in aggression and in response to the complex environment. However, if these results are confirmed there will be some difficult ethical choices between environmental enrichment on the one hand and the use of more animals in each experiment on the other.

3

Understanding and controlling variation

Introduction

Experiments can only be designed efficiently and economically if the research worker has a good understanding of biological variability, its causes, and ways in which experiments can be arranged in order to quantify and control it. There are two types of variability that need to be taken into account: random variability (or random effects) due, for example, to uncontrolled inter-individual differences; and 'fixed effects'—such as the sex, strain, age, diet and bedding—which can be controlled to a large extent by the investigator. The ways in which these are handled are discussed in this chapter.

The majority of statistical tests are basically comparing the size of the effect (the biological 'signal') relative to the amount of variability in the data (the 'noise'). Figure 3.1 illustrates what might be concluded about the same biological effect, under two different scenarios. In the first of these, the background variability is relatively low, and in the second the variability is relatively high.

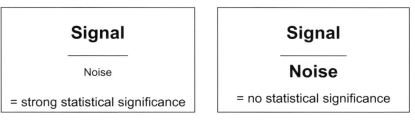

Figure 3.1 If the noise is low then the signal is detectable, but if the noise (i.e. inter-individual variation) is large then the same signal will not be detected

An analogy would be a lecturer trying to speak while a radio played loudly in the same room. The lecturer has a number of options:

◆ Speak louder *(increase the signal)*
◆ Turn the radio down *(decrease the noise)*
◆ Find a different way to get the message across *(e.g. seek an alternative signal)*

In a statistical test, the ratio of signal to noise determines the significance. Hence if the variation or 'noise' is large in an experiment, the biological effect, the 'signal', may be hidden by it. There are usually many sources of noise in biological data. Imagine the same lecturer in a room with many radios. The task of turning the radio down should now involve a first step of identifying the loudest radio(s). Only then can the noise be reduced. The experimenter should not automatically assume that the animal is the main source of noise and reduce it by increasing animal numbers. Measurement error of various sorts may well be more important, and this can often be controlled by better experimental technique, by increasing the number of determinations, or by using covariance or blocking to eliminate otherwise uncontrollable variation.

Sources and types of variation: fixed and random effects

What causes the variability? In a training course, a group of scientists were asked to write down possible sources of variability in their experiments, and these were grouped together into the categories shown in Table 3.1. These variables do not all, necessarily, cause experimental noise. They fall into two main classes designated 'fixed effects' and 'random effects'.

Fixed effects

These variables may affect the outcome of an experiment, but they are largely under the control of the experimenter, and are of

Table 3.1   Some variables that may influence the outcome of an experiment

Environment	Temperature, humidity, season, barometric pressure, lunar cycle, noise, air movement, light, smells, room characteristics, cage size and design, bedding material, nest box design, nest materials, number of animals in group, water quality, diet type, diet availability, diet quality, frequency and duration of handling
Animals	Species, sex, strain, genotype, health status, batch, supplier, age, body weight, litter size, oestrus stage of females, level of inter-animal aggression
People	Researchers, surgeons, animal technicians, observers, passers by
Experimental	Type/quality of surgery, route of injections, dose levels, sampling of tissues and/or fluids, time (day-to-day, hour-to-hour, etc), variation in test materials, preparation of test materials, shelf-life of solutions, calibration of instruments, measurement errors, recording errors

particular importance when considering the design of the experiment and when interpreting the results. The imposed treatment is the most obvious fixed effect, but the species, strain, sex, age or weight range of the animals, type of caging, type of bedding material and many of the other factors listed in Table 3.1 can be specified by the researcher. Strictly, it is a matter of scientific judgement whether the conclusions from an experiment using, say, male rats can be generalized to female rats or whether it makes any difference whether the rats were eight or ten weeks old at the beginning of the experiment. But if it is expected that the response may be different in the two sexes, or age groups, then there are two possible courses of action. First, the scientist may be content to state that the results are only applicable to the sex or age of animals used in the experiment. Second, a factorial experiment (see Chapter 4) involving both sexes or ages as well as the treatments of interest could be used to explore whether or not this is the case.

All experiments involve the choice of a wide range of fixed effects. It is up to the research worker to decide which effects are likely to be of trivial importance, which may influence the interpretation of the results, and those which might need to be explored using factorial experimental designs.

Random effects

These are the variables that usually contribute 'noise' or unwanted variability among the experimental units. Variability in body weight within the specified weight range, genotype within an outbred stock (although the experimenter can choose whether to use Wistar or Long-Evans rats, so the stock is usually a fixed effect), accidents of development, social hierarchy and within-group aggression, subclinical variation in pathogen burden, poorly mixed diet, inaccurately administered dose levels, contamination of blood or urine samples, inaccurate measurements or measurements made near the limits of detectability all contribute to variability and noise.

In an experiment, random effects are nearly always to be controlled or avoided, but this is done in a different way from fixed effects.

◆ The first step is to try to obtain experimental material with a low intrinsic noise level. Isogenic animals (see Chapter 2), free of pathogens and of a narrow age/weight range, well acclimatized and housed in compatible groups in a good animal house will often be the starting point.

◆ Variability that occurs over time due to biological rhythms or the shelf-life of reagents can often be partially controlled by blocking or covariance (see Chapter 4), as can some of the variability due to position of the cages in the animal house.

◆ Measurement errors can often be substantially reduced using multiple sample determinations. There may be a whole hierarchy of possibilities. In isolating enzymes or mRNA from a liver, several samples of liver might be taken and several determinations could be done on each. Exactly how many liver samples and determinations per sample are chosen could largely determine precision.

◆ Finally, if unacceptable levels of variability persist, and cannot be identified and controlled, then sample sizes can be increased to give an experiment with sufficient statistical power (see Chapter 5).

The distinction between fixed and random effects is not always clear-cut. Body weight is a fixed effect if a weight range (e.g. 100–120 g) is specified, but within that range it will be a random effect that may cause noise. Likewise, if unsexed animals are used, then sex would be a random effect, but more usually the sex of the animals will be controlled, making it a fixed effect. The important point is whether the variable contributes uncontrolled noise, or whether it can be specified and controlled and/or studied as a factor in a factorial experimental design.

Although the treatment is usually considered to be a fixed effect, in some cases an experiment is used to determine the magnitude of a random effect. For example, the effect of variation in commercial mouse diets on body weight of mice maintained on the diets could be determined by taking a random sample of diets from all available diets, and feeding these to mice under specified conditions. The results would normally be expressed in terms of the proportion of variation in mouse weight that could be attributed to variation among the diets. An example of this type of experiment and the associated statistical analysis are given in Chapter 4.

Identification of the important sources of variability

If the variation associated with what are assumed to be the most important random and fixed effects can be quantified, then a rational approach can be used to control or minimise its impact. For

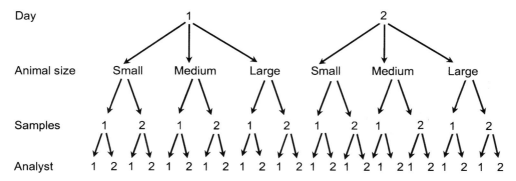

Figure 3.2 Structure of a 'nested' experiment designed to quantify the amount of variation between days, sizes of animals, samples, and analysts. A design of this sort quantifying variation between animals and between microscopic fields of sections of the hippocampus is described in more detail in Chapter 4

example, a randomized block design may be used if there are large fluctuations in mean results over a period of time or if the experiment has to be split into two rooms which may have different environments. Pilot experiments can be useful if several similar experiments are contemplated. For example, the experiment shown in Figure 3.2 was designed to explore the importance of day-to-day variation, animal size, the number of samples taken, and the analyst who analyses the samples. Note that animal size is a fixed effect, but the other variables are random effects if it is assumed that the two analysts are chosen from a pool of possible analysts. The resulting data could be analysed to show the relative importance of each of these variables. Thus, if there were large differences associated with animal size, then it would imply that body weight would need to be rigidly controlled and possibly included as a factor in a factorial experimental design. The latter would be appropriate if there were a suspicion that the response might be different in large and small animals. If the day-to-day variation was relatively large, it could be controlled using blocking (see Chapter 4). If there were larger differences between samples, then triplicate or quadruplicate samples might be appropriate.

Some examples

Atherosclerosis

In a study to investigate the effects of drug treatments in a rabbit model of atherosclerosis, there were three drug treatments and four animals on each drug. At autopsy the aorta was removed and five lateral sections were cut. The lesion area in each section was measured using image analysis software. The layout is somewhat similar to that shown in Figure 3.2, though in this case only three levels of nesting were involved. An hierarchical analysis of variance (discussed in Chapter 4) was used to quantify the variability between sections, animals and drug treatments, and it was found that 47% of the variation was due to the variation between sections, 35% due to the effect of the drug (a fixed effect), and 18% due to the inter-animal variation. Thus, the best way of improving the power or precision of such an experiment would probably be to take more sections, rather than using more animals, especially as this may also be the cheapest alternative both financially and in terms of animal use.

Stroke

In a pilot experiment to study the effect of a new drug on a model of stroke caused by occlusion of the median carotid artery in rats, a factorial design was used to explore the effect of the three occlusion periods (30 min, 45 min and 60 min) and two drugs, using five rats per group. Note that the occlusion period and drug treatment are fixed effects, hence the use of a factorial design (see Chapter 4). Most responses were highly variable. However, it was found that the variability was minimal with the 60-minute occlusion period. With the 30-minute period some rats were not developing stroke lesions, hence they were not responding to the drugs. Thus, rather than increasing the number of animals, it would be better to use the longer occlusion period in order to design more powerful experiments in the future. Factorial designs are often used in this way to determine an optimum set of conditions for obtaining the maximum yield of a product.

Randomization

'Randomisation is one of the essential features of most experiments. The investigator who declines to randomise is digging a hole for himself, and he cannot expect the statistician to provide the ladder that will help him out.' (Finney 1970)

The aim of experimental design is, as far as possible, to remove all sources of difference from the experimental units except for the explicit treatment or intervention. Although many differences can be controlled, some variation will always remain. However similar they may be, no two animals are identical and even if they were identical it would be impossible for the same operator to administer the treatment simultaneously, thus giving rise to differences in operator and time. Randomization is essential to ensure that these remaining and inescapable differences are spread among all treatment groups with equal probability, thereby providing a reliable estimate of experimental variation or error and minimizing any potential bias.

It follows from this that 'control' groups, where one of the treatments is considered as a control, cannot be separated from the experiment. 'Historical controls', for example, are unlikely to be valid. Valid use of a control group requires that the 'control treatment' be one of the treatments in the experiment. 'Blinding' the experimenter to the precise nature of the treatments, including the controls, should as far as possible be included in the randomization procedure — that is, the experimenter should not know which animals have received a particular treatment, in order to avoid any possible bias.

Basic randomization procedures

Use of a computer

It is important to use an objective method. A number of statistical packages can be used to assign units and treatments randomly to a specified design. For example, in MINITAB (Minitab Inc., 3081 Enterprise Drive, State College, PA 16801-3008, USA) the numbers 1–20, say, in column one of the data sheet can be put into column two in a random order with a simple menu-driven command. Thus if 20 rats are to be assigned at random to four treatment groups, the rats will be numbered 1–20 and those with the numbers shown in the first five rows of column two would be assigned to treatment one, and so on. When this was actually done, rats 14, 15, 20, 9 and 8 were assigned to treatment one, rats 3, 4, 18, 7 and 19 to treatment two and so on.

A spreadsheet such as EXCEL can be used for randomizing animals. If the command $=TRUNC(RAND()*n,0)$, where n is the highest number that is wanted, is put in a cell it will return a digit between 0 and n. This cell can be copied and pasted to produce a column of numbers. Thus, if the aim were to randomize 20 rats to four treatment groups, the rats would first be numbered 1–20. A random column of about 50 numbers, between 0 and 20 could be generated and the first five would be assigned to treatment one, etc. More than 20 numbers are needed to allow for repeats. This whole procedure could be done on the computer before the animals are numbered and assigned in the animal house.

Physical randomization

Physical randomization is extremely easy. The experimental units (e.g. animals) are numbered, and the numbers are also written on slips of paper which are folded and put in a receptacle. This is thoroughly shaken (not stirred!), and then the first, say, five numbers drawn are assigned to treatment one, the next five to treatment two, and so on.

Tables of random numbers

Tables of random numbers provide another easy method without recourse to statistical packages. These are printed in most statistical textbooks. Suppose that 10 mice need to be put into two groups of five. The mice should be numbered 01, 02, etc. to 10. Now, without looking, a pin is placed on the printed page of random numbers to hit a random digit. This digit is used to specify the row and the next one to the right, a column. The digit specified by the row and column is read. This might be, for example, 84. However, the only useful numbers are those between 1 and 10. So numbers are read off down the column until such a number is found. If necessary, the next column can be used. Suppose the number 5 is found, then mouse number 5 is assigned to treatment group A. Once five mice have been assigned to group A, the remaining mice are allocated to group B so that the group sizes are equal.

With three treatment groups and, say, 15 mice to be allocated, the first group is filled with five mice, as above, then the second group, and the remaining mice are then assigned to the third group.

Randomization of a randomized block design

There are many instances when the 'cage' or some other characteristic of the experimental units may be used as a blocking factor. However, this does not eliminate the need for randomization. For example, in an experiment with three treatments, drug 'A', drug 'B', and control 'C', using 18 animals, nine of the animals may be male and the other nine female. Cage 1 will contain nine females, with three receiving drug A, three receiving drug B and three controls; and Cage 2 will contain nine males, similarly assigned to three groups.

Treatments should be assigned randomly to codes X, Y and Z. For example, the controls might be designated treatment Y. These three treatments should then be randomly assigned to the animals in the cage. The females in Cage 1 should be numbered from 1 to 9. These numbers should then be selected from a table of random numbers as indicated above, or using physical randomization, and animals with the first three numbers should be given treatment X, those with the next three numbers treatment Y and those with the last three numbers treatment Z. In this design, the blocking factor is 'sex and cage' (which cannot be separately considered) and there is within-block replication. This design would thus not be appropriate if the experimenter wished to determine whether there were differences in the way the sexes responded to the treatments.

'Improving' on the randomization

Some people advocate 'improving' on the random allocation of animals to the treatment groups by moving animals from one group to another so as to have exactly the same mean body weight in each group before starting the experiment. The problem with this approach is that by minimizing the differences between the groups, the variation within groups is increased. This may result in a reduction in the power of the experiment. In any case, body weight is not the only variable to be considered when setting up the experiment, so this approach is not recommended.

'In short, randomisation provides an assurance, not only to the experimenter, but to others who may be more sceptical than he, that the magnitude of the ordinary sources of disturbance ... has been evaluated by means of the estimate of error. It does not, of course, provide a panacea which removes all need for care and foresight; it cannot take account of types of disturbance which act selectively on the various treatments, ... and a badly planned or carelessly executed experiment will still be inaccurate even though it is randomised, but the experimenter will at least know of its inaccuracy.' (Yates 1939)

4

The design of experiments

Introduction

An experiment makes it possible to infer whether or not a treatment *causes* changes in a variable or outcome of interest. Statistical methods are used to calculate the probability of error in the process of decision making. Even if the effect of the treatment is very clear cut, it is likely that there will be some 'noise', 'error' or residual variation in the results, as outlined in Chapter 3.

An experiment should not be confused with a survey, which merely gathers information about a situation or process that cannot be controlled. In a survey, the variable of interest (e.g. levels of smoking) cannot be deliberately varied, and inferences about its effects (e.g. on health) in a population can only be based on associations or correlations, but not on causation. Thus, strictly, surveys cannot show that smoking causes lung cancer in humans, only that it is associated with lung cancer. However, this particular association is so well established, and there is such good evidence that tobacco smoke contains compounds which cause cancer in animals, that most people are now convinced that it could only have arisen by causation.

The purpose of experimental design, and of this chapter, is to provide guidelines on data collection in such a way that causation may be established beyond a reasonable doubt and the residual variation minimized so that most of the variation is explained by the variables that are deliberately manipulated. The designs may also need to incorporate some extra considerations, such as the efficient use of more than one sex or strain, so that the interpretation of the results may make it possible to draw wider conclusions about the underlying science. A summary of the designs most commonly used in biomedical research is given in Table 4.1, with more detailed descriptions given later in this chapter.

Table 4.1 Summary of the most common experimental designs reviewed in this chapter

Type of design	When to use	Advantage	Disadvantage	Notes
The completely randomized design (fixed effects; random effects)	Fixed effects: when uncontrollable sources of variation are unlikely to be important Random effects: when only interested in quantifying source of random variation (e.g. surveys) Useful for within-subject studies rather than compare treatments.	Simplest 'designs' Easy to use and analyse Less affected by unequal sample size	No control of additional 'nuisance' variation caused by uncontrolled time and space variation etc. which might affect results	Typical designs used when doing simple t-tests
Block designs (including Latin square)	When wanting to remove the effects of known or suspected 'nuisance' variation. The evidence shows that it is under-used in the biomedical literature, particularly in its simplest form (controlling at least one source of unwanted variation)	Deals with known or suspected 'nuisance' variation in a systematic and powerful way by breaking the experiment into a series of subunits which are analysed as a whole. Un-equal sample sizes present some problems	Can be sensitive to missing values Becomes increasingly complex when dealing with two (Latin square) or three sources (Graeco-Latin squares) of unwanted variation	These designs aim to improve the *precision* of the experiment. They can be used in conjunction with others (e.g. factorial designs). Analysis of covariance, another way of increasing precision is also discussed
Factorial designs	When wanting to increase the 'generality' of the results, by testing *simultaneously* the potential effect of several factors (e.g. treatment, sex, strain) and their interactions on the response (dependent variable[s])	A much more powerful alternative to doing several smaller experiments for each factor. Allows testing for interactions between these factors	Interaction between more than two factors are often difficult to interpret	These designs aim to increase the *amount of information* yielded by the experiment. They can be used in conjunction with other (e.g. blocking) designs
Repeated measure designs (crossover, split plot, mixed effects designs)	When using several measures on the same individual, either to study change over time or as a way of dealing with strong individual variation	A more powerful alternative to the dreaded repeated t-tests! The individual can act as its own control (crossover)	Crossover: not valid if there is a strong order effect. It may be dangerous to consider time as an independent variable in these experiments	High precision as a result of eliminating inter-individual variation
Sequential designs	When results with individual animals can be obtained quickly, and where the aim is to minimise animal use	Results are analysed as the experiment unravels, enabling the experimenter to stop as soon as significance is obtained	Requires expert advice. Logistics may be a problem	Up-and-down method may be replacing the classical LD50 test in the USA

Principles of design and analysis: some reminders

Good experimental design relies on two principles:

Replication: The more times something is repeated, the greater the confidence of ending up with a genuine result. For example, with a choice of two routes to drive into work, each of about the same distance, a person would usually want to drive each route more than once before being happy to choose one route over another. However, this concept cannot be used as a justification for running very large experiments. As explained in Chapter 3, replication should be across the source of variation which is likely to be the biggest cause of variability in the experiment. The determination of sample size is discussed in more detail in Chapter 5. Replication also makes it possible to detect abnormal observations or outliers, perhaps due to errors of measurement or recording. Statistical analysis is used to make estimates of the probability of error in reaching conclusions from the data, and in particular to prevent claims of an important treatment effect when the results could well be due to sampling variation.

Randomization: As outlined in Chapter 3, experimental subjects (e.g. animals) must be allocated to treatment groups at random, and the order in which subjects are tested must be at random in order to avoid biases that would be almost certain to occur with any other method of allocation.

The analysis of variance (ANOVA)

There is a close relationship between the design of an experiment and the statistical methods that will be used to analyse the resulting data. The 'analysis of variance' (ANOVA for short) is a highly versatile method of analysing quantitative data, which can be used to analyse very simple experiments such as those involving a comparison of the means of two groups (in which case it gives identical results to Student's *t-test*), as well as more complex designs such as randomized blocks, repeated measures and factorial designs, which are discussed later in this chapter. It is almost essential for anyone doing experiments on animals to have a basic understanding of this powerful method, because it is often the only method that can be used for analysing many of the designs described here.

The ANOVA divides the total variation in the dependent variable into parts, shown diagrammatically in Figure 4.1. Note that 'variation' is quantified as sums of squared deviations from group or overall means. First, the overall mean across all treatment groups

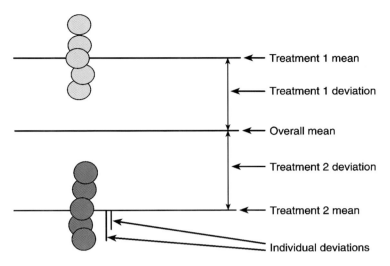

Treatment 1 mean

Treatment 1 deviation

Overall mean

Treatment 2 deviation

Treatment 2 mean

Individual deviations

Figure 4.1 Diagram showing treatment deviations from the overall mean, and individual deviations from the treatment means used in the analysis of variance. The treatment sum of squares is the sum of the squared deviations of the treatments from the overall mean, and the error sum of squares is the sum of the squared deviations of each observation from its treatment mean

is calculated, then the deviations of each treatment group from the overall mean are quantified by adding up the squares of these deviations (the Treatment 'Sums of Squares'). Finally, the remaining unexplained ('error') deviations of each individual data point from its treatment mean are squared and added up to give the Error Sums of Squares.

The ANOVA produces a summary table similar to the example given in Table 4.3 (*see* page 48). The column headed 'Source' indicates the source of the variation, which usually consists of treatment (in this case labelled 'running'), error and total variation, but may contain other sources of variation such as blocks, rows and columns (say in a Latin square design) and covariates (in the analysis of co-variance). DF stands for the *degrees of freedom* for each source of variation, i.e. the number of bits of information available on the particular source of variation. It takes two observations to obtain one bit of information about variation (estimated from the difference between them), and N observations will generate N−1 bits of information about variation among objects. Thus if there are five treatments, there will be 5−1 = 4 DF for treatments. The number of DF for the error or residual variation is usually estimated by subtraction, but is done automatically by the computer package.

The column headed SS gives the sum of the squared deviations from the respective means. These quantify the variation associated with each source. These are converted to the MS (mean squares) by dividing them by the DF.

The column headed F shows the 'variance ratio' statistic, being the treatment MS divided by the Error MS, and designated as 'F' in honour of R A Fisher who developed the ANOVA method. Finally, P is the probability that an F statistic as large as or greater than the calculated one could have arisen by chance, in the absence of that particular source of variation (e.g. some treatments) being present. More commonly it is described as *the probability that treatment means at least as disparate as observed in this study could have arisen by chance* (i.e. the chance that it could merely be due to sampling variation in the absence of any true treatment effect). By convention, a P value of >0.05 is often considered to be 'not significant', one of <0.05 is considered 'significant' and one of <0.01 is 'highly significant'. However, these cut-off points are entirely arbitrary and should not be interpreted too rigidly. A P value of 0.06 does not necessarily mean that the treatment has had no effect. It might just be that the experiment was too small or the variation among individuals too great to be able to detect an effect.

Assumptions of the analysis of variance

The validity of the ANOVA depends on three assumptions, which should normally be examined critically as part of the statistical analysis:

(1)   The residuals (i.e. the deviation of each observation from its group mean, see Figure 4.1) should have a normal distribution. Note that it is the residuals that must be normally distributed, not the data themselves.

(2)   The variances within each group should be approximately equal.

(3)   The observations are independent of one another.

Most good statistical packages offer 'residuals diagnostic plots' which aim to show graphically whether the first two of these assumptions are valid. The validity of the third assumption will depend on exactly how the experiment was conducted, whether it was appropriately randomized, and whether the appropriate experimental unit was chosen. The use of residuals plots to evaluate the assumptions is discussed later in this chapter.

Comparisons of treatment groups following the ANOVA

When several treatments are being compared, the ANOVA indicates whether there is good evidence that the treatments are different.

However, with three or more treatment groups a scientist usually wants to know which groups differ from the controls, or from each other, or whether there is a trend in response according to dose level, etc., depending on the exact nature of the study.

There are two main ways of making such comparisons. The most efficient is to plan exactly which comparisons are of interest, and use an appropriate set of what are called 'orthogonal comparisons'. This is a fairly flexible method that can be used, for example, to study a range of comparisons of treatment means. The method can also determine whether there is a linear and/or curved trend in the response to dose levels of a compound. Unfortunately, the methods are only supported by the more advanced statistical packages and although the calculations are not difficult, and can be done by hand, it is not always easy to understand from the text books how they are done. More details are given by Altman (1991) and Snedecor & Cochran (1980).

Another possibility is to perform *post-hoc* ('after the event') multiple comparisons of treatments. These all work on the premise that the more statistical tests are done on the same set of data, the greater the risk that at least one of these tests will be significant purely by chance (a false-positive result). Multiple comparison tests control the risk of false-positive results taking the experiment as a whole. The more comparisons that are made, the greater the penalty, which obviously can reduce the sensitivity of the test to pick up a genuine treatment difference as statistically significant (a false-negative result). There are several multiple comparison tests. These are discussed in most statistical textbooks (Altman 1991, Dixon & Massey 1983, Maxwell & Delaney 1989, Petrie & Watson 1999, Snedecor & Cochran 1980). The following tests are commonly used:

(1) Methods for comparing several treatments include Fisher's Least Significant Difference (LSD), Scheffe's test, Duncan's multiple range test, Neuman-Keuls' test, Tukey's test.

(2) Dunnett's test is appropriate for comparing several dose groups with a control group.

(3) Bonferroni's method can be used for making a small subset of treatment comparisons, with the P value for each treatment comparison being declared significant if it is less than α/k, where α is the chosen significance level (say 0.05) and k is the number of pair-wise treatment comparisons. This method can lead to too many false-negative results as k gets too large.

These methods should only be used if there is a significant overall treatment difference in the ANOVA. In general, experiments should be designed to test relatively simple hypotheses, and large numbers of *post-hoc* comparisons should be avoided.

Considerations in the choice of design

As already noted, Table 4.1 summarizes some of the more common experimental designs used in biology. All researchers should be aware of these before embarking on experiments with animals.

Starting with the very simplest design (a completely randomized, single-factor design), there have been two major ways in which more complex designs have developed:

(1) **Designs for improving precision:** The precision of an experiment is increased by better control of variation and/or by taking account of some natural structure in the experimental material. Randomized block and Latin square designs eliminate some sources of heterogeneity by choosing subsets of uniform experimental units which are then assigned at random to the treatment groups. Crossover designs involve sequential experimentation on the same subjects, which should eliminate some inter-individual variation.

(2) **Designs for increasing the amount of information from each experiment:** Factorial designs, which involve two or more independent variables such as some treatments and both sexes, or several treatments and termination times, will usually produce more information without necessarily increasing the size of the experiment. The aim of these designs is to find out the effects of each factor separately, and their joint effects. Advanced versions of these designs can be used, for example, to find the optimum combination of a large number of factors. For a given input of resources (animals, reagents, time, etc.), factorial designs will normally provide more information than a single factor design, at little or no extra cost. There seems to be a widespread misunderstanding among scientists that if an experiment needs, say, 24 male mice, then an experiment involving both males and females will need 48 mice. This is not true. In many cases the experiment involving 24 male mice could be done using 12 male and 12 female mice (a factorial design) with little loss of precision, and with a useful increase in the amount of scientific information produced.

Fixed and random effects models

Usually the experimenter chooses the treatments to be tested and is interested in the differences between the means of these treatments. The experiment could be repeated subsequently under the same conditions with the same 'fixed' set of treatments. Treatments may be drugs, strains, diets, methods, etc., and the 'controls' are considered to be one of the treatments, even if they are untreated. These treatments are fixed by the experimenter, and the resulting ANOVA model is known as a 'fixed effects' model.

An alternative situation is when the 'treatments' can be considered as a random sample taken from a population of possible 'treatments' (treatments are given here in inverted commas because often they are not what most experimenters would recognize as a treatment). An example here would be testing over a number of days, or perhaps between a number of batches of animals, or batches of diet, or over a range of body weights, where days, batches or body weight groups are regarded as the 'treatments'. While there may be little interest in the effects of the specific random sample of treatments, there may be some interest in learning something about the level of variability inherent among batches, etc. For example, if there was a large amount of variability due to the time or day of testing, then this could explain many of the problems encountered in repeating this type of study in the future. This unwanted variability can be estimated in order to determine how much effort should be put into controlling it. This model is known as a 'random effects' model.

There is a third type, the 'mixed model', which combines fixed and random effects. Strictly, the randomized block experimental design is a mixed model as the treatments are fixed and the blocks are a random effect. However, there is usually little interest in estimating the magnitude of a block effect, whereas in some other cases part of the aim of the experiment may be to estimate the amount of variation associated with the random effect.

An example

Perhaps the best way to illustrate the design options open to the experimenter is to start the way that most research projects begin—with a question. Does jogging improve memory? Suppose that the aim is to test the hypothesis that running has an effect on learning and memory. Contrary to the old belief that the number of neurons can only decrease after birth, it is now known that new neurons can be added to the adult brain. In some bird species, the area of the brain associated with singing grows during the breeding season, and shrinks after that (Tramontin & Brenowitz 2000). In mice, exposure to an enriched environment increases the production of new neurons (neurogenesis) in some parts of the brain (Kempermann et al. 1997) and so does general physical activity.

Suppose, therefore, that an experiment is planned to test whether physical activity (e.g. wheel running) enhances learning ability and memory in mice (see Van Praag *et al.* 1999). How should it be designed? In many ways this depends upon the questions being asked, i.e. what should be measured, how much additional ('nuisance') variation is likely to be caused by, say, sex, genotype or weight, that may mask the effects, and are there interactions between the independent variables that are used? The first point about design is that the experimenters must be very clear about the hypotheses to be tested.

The completely randomized design

With this design, experimental units are allocated to treatments completely at random, though sometimes subject to the restriction that there should be equal numbers in each treatment group, hence the name. There is only one factor ('the treatment') with any number of treatment groups or levels, and any number of replicates per treatment. These designs have the advantage of being easy to use, they are relatively unaffected by unequal numbers in each treatment group, and are easy to analyse. But they do not specifically control any inherent variability in the experimental units (e.g. weight, genotype, etc.) and may therefore be inefficient if the experimental material is heterogeneous. Nevertheless they are useful to introduce basic principles and assumptions.

In an effort to test whether running (described as the explanatory variable, independent variable or treatment) affects learning (the response or dependent variable), three levels of running activity in a running wheel might be chosen:

(1) no running (a non-rotating wheel is placed in the cage, say for 30 minutes)
(2) moderate running (mice are allowed access to a running wheel for 30 minutes per 24 hours)
(3) marathon running (access to a running wheel for three hours per 24 hours).

These treatments are to last for three weeks before the mice are tested for learning ability in a maze. The level of learning ability constitutes the response (or Y variable). There are many ways that learning ability can be measured in practice. For simplicity, the response is designated 'learning ability' without detailing the specific measure that was used to determine it. Low values represent good learning ability.

Formally, the hypotheses of interest are:

H_0 (**the null hypothesis**)—all means are equal (i.e. there is no effect of running activity on learning ability)

H_1 **(the alternative hypothesis)**—at least one mean differs from the others (i.e. running has some effect).

The data comprise nine observations for each running treatment, and are shown in Table 4.2. Note that there are two kinds of variation: between and within running regimes. The former refers to the deviation of the treatment mean levels from the grand mean; the latter the deviation of each observation from its corresponding treatment mean ('noise'). If the null hypothesis is true, the ratio of between to within variances should be small: this ratio is the F value from the ANOVA table.

Data screening

Before commencing a formal statistical analysis, it is recommended that the data are screened for obvious errors and departures from normality. For example, a preliminary ANOVA to produce residuals plots (but without looking at the ANOVA results at this stage) will often show outliers or serious deviations from normality. If there are two dependent variables which are correlated, they can be plotted against each other. This will often detect outliers. The observations may also be plotted against the treatment codes. This step identifies any unusual data points or trends in the data which may inflate the variability and otherwise mask a genuine treatment effect. Outliers should be investigated in case they have arisen as a transcription error, but should not be discarded. It may also make it possible to:

(1) choose an appropriate data transformation if the data are skewed or if the variability changes with the size of the mean response;
(2) choose an appropriate summary measure for repeated measures data;
(3) identify a possible covariate (to be discussed in a later section);
(4) interpret the results of the statistical analysis in a more meaningful way;
(5) decide that an ANOVA is not going to be suitable for that set of data, so that an alternative non-parametric test can be chosen.

Table 4.2 Individual and mean learning ability from three wheel-running treatments

	Observations									Mean
No running	238	250	246	258	251	259	252	230	231	246.11
Moderate running	216	241	227	228	238	229	242	212	221	228.22
Marathon running	233	212	219	229	218	205	218	232	230	221.78

These are not discussed here, but are covered in most statistical texts or specialized monographs (Maxwell & Delaney 1989, Petrie & Watson 1999, Sprent 1993).

Analysis

These data are analysed with a one-way ANOVA, as only a single factor (running) is involved. This is easily done using statistical software. The ANOVA table is shown in Table 4.3. Clearly, the effect of running on memory is significant. ($F_{2,24} = 13.13$; $P < 0.001$).

Planned comparisons

In order to establish which treatments differ, using the SPSS statistical package, two orthogonal contrasts can be constructed for the differences between the controls and running (moderate and marathon averaged) and between moderate and marathon running (Table 4.4). The calculations have also been done by hand in Table 4.4. The variation caused by the treatment, running, is the same as that found previously. Note that the DF and SS of the two contrasts add up to the overall one caused by running.

Multiple comparisons

Table 4.5 shows an example of output from the SPSS statistical package using the Bonferroni method. When the analysis is done with MINITAB using Tukey's test, the confidence intervals are slightly different (not shown), but the conclusions are essentially the same. There is no significant difference ($P > 0.05$) between the two treatments involving some running, though the graph does show that the marathon running yields significantly ($P < 0.05$) lower learning times than no running. A larger sample size might be necessary to establish whether the difference between no running and moderate running is significant.

Table 4.3   Analysis of variance of the wheel-running experiment and learning ability. Note that a *P* value of 0.000 does not mean that it is zero, only that it is less than 0.0005

Source	DF	SS	MS	F	P
Running	2	2861	1430.5	13.13	0.000
Error	24	2614	108.9		
Total	26	5475			

Table 4.4 Analysis of variance with two orthogonal treatment contrasts. These were calculated long-hand using the formulae given in Snedecor & Cochran (1980)

Source	DF	SS	MS	F	P
Running	2	2861	1431	13.13	0.000
C vs. (M + M) (1)	1	2674	2674	24.55	0.000
(M vs. M) (2)	1	187	187	1.72	>0.05
Error	24	2614	109		
Total	26	5475			

(1) Comparison of the mean of the controls versus the means of the two running groups (Moderate and Marathon)
(2) Comparison of Moderate versus Marathon running

Note: The calculation of the sums of squares for orthogonal contrasts does not involve any tedious calculations, though understanding how it is done may be a problem. For this example, the calculations are set out below: Briefly, a 'contrast' (C) is a difference between means or groups of means. Thus, the mean of the moderate group (M1) is 228.22 and of the marathon group (M2) it is 221.78, so $1 \times 228.22 - 1 \times 221.78 = 6.44$ is a treatment contrast (C1). Note that the coefficients (1 and −1 in this case) involved in a contrast must equal 0. Similarly $2 \times$ (control mean) $- (1 \times M1 + 1 \times M2)$ is another contrast, C2. Note that $2 - (1 + 1) = 0$. Two contrasts are said to be orthogonal if the product of their coefficients equals zero. In this case cross-multiplying the coefficients of C1 and C2 gives $(2 \times 0) + (1 \times -1) + (-1 \times -1) = 0$. So C1 and C2 are orthogonal contrasts. Now $SSC = NC^2/(sum$ of coefficients squared), where SSC is the sum of squares for a contrast, N is the number making up each mean and C is the contrast. Thus SSC, $9 \times 6.44^2/2 = 186.63$ is the sum of squares for C1 and SSC2 = $9 \times (2 \times 246.11 - 228.22 - 221.78)2/6 = 2673.79$. Note that the divisor six in this case is made up of $2^2 + 1^2 + 1^2$, the sums of squares of the coefficients

Table 4.5 Wheel running/learning ability experiment. Output using the Bonferroni correction method (SPSS). The output shows the 95% confidence intervals for specified linear combinations (i.e. differences between means) by the Bonferroni method. Note that if the Lower Bound and Upper Bound for the differences between the means exclude zero, then the means are significantly different
Critical point: 2.5736
Response variable: learning
Intervals excluding 0 are flagged by****

	Estimate	Std. error	Lower Bound	Upper Bound
No running − Moderate running	17.90	4.92	5.23	30.6****
No running − Marathon running	24.30	4.92	11.70	37.0****
Moderate running − Marathon running	6.44	4.92	−6.22	19.1

Testing assumptions

As already mentioned, this analysis requires three assumptions: homogeneity of variances, normality of the residuals, and independence of errors:

Homogeneity of variances

The variation should be approximately the same in all groups. Plots of fits versus residuals can be used to check this assumption. All good statistical packages provide such plots. Figure 4.2 shows this type of plot for the running experiment. The spread of the observations appears to be about the same for each of the three

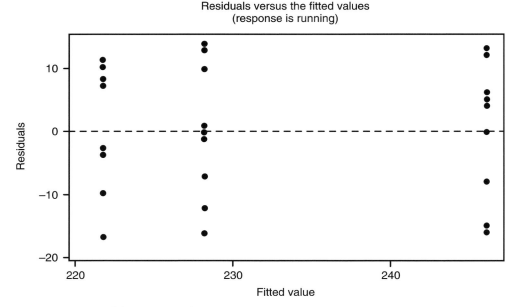

Figure 4.2 Plot of fitted value (group means) versus residuals (deviations from group means) for assessing whether the variation is similar in each group. Note that the variation appears to be very similar in all three groups

groups, so this assumption appears to be valid. If in doubt, Bartlett's or Levine's test can be used to test this assumption formally, though in most cases the plots are good enough. The ANOVA is quite robust so that some departures from the assumptions will have relatively little effect.

Normality

Most computer packages also provide a normal plot of the residuals as part of the ANOVA procedure. If the residuals are normally distributed, then the plot will give a straight line. For the running example, this is shown in Figure 4.3. There is some slight curvature, apparently due to truncation of very low and very high scores, but this is probably of no great importance. A formal test for departures from normality can be used, also available in many computer packages, though this is rarely necessary. In this case, MINITAB provides a formal test of normality, and the deviation from normality is not statistically significant $(P = 0.178)$.

Independence of errors

There is no formal test for independence of the errors. This depends on the design of the experiment and the way in which the

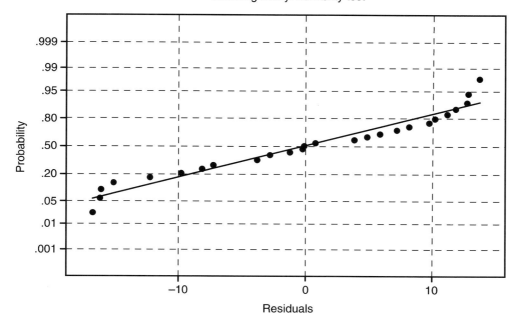

Figure 4.3 Normal probability plot of residuals from the running experiment. If the
residuals have a normal distribution, this should be a straight line. This is a plot from the
MINITAB test for normality. In this case there is some evidence of non-linearity at the
extreme ends, but the null hypothesis that there is no departure from normality cannot be
rejected at the 5% significance level ($P=0.178$)

observations were collected. Assuming the mice were assigned to
the treatment groups at random, that the cages were in random
order on the shelf, and that when the learning ability tests were
conducted they were done blind and in a random order, then the
errors should be independent. Thus, each mouse's response should
be independent of that of the other mice.

Exploring variability through nested designs

Sometimes, the aim may be to quantify the sources of random
variation, rather than compare means. For example, previous work
may suggest that the density of the neurones in the hippocampus
may be affected by wheel running. A pilot study to find out how
much variation there is in the density of the neurones in this region,
both between mice and between microscope fields, might be
planned to find out how many mice and how many fields would

need to be studied. This might involve a sample of, say, five mice and for each mouse four randomly chosen microscopic fields of the hippocampus, with the number of neurons being counted in each field. The 20 observations (5 mice × 4 samples, see Table 4.6 & Figure 4.4) provide information on the between- and within-mouse variability.

Table 4.6   Data for a components of variance model analysis

Mouse	Density of neurones in each of the four microscope fields			
1	187	162	162	178
2	165	149	156	150
3	154	140	151	142
4	159	162	171	140
5	170	164	156	151

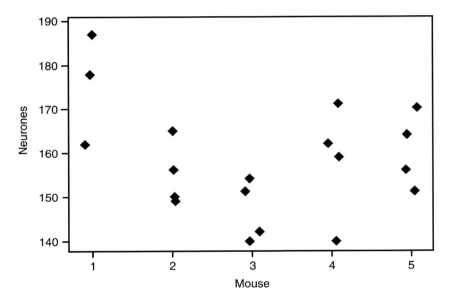

Figure 4.4 Count of neurones in four microscopic fields of the hippocampus in five individual mice. Data given in Table 4.6. Note that some 'jitter' has been added on the X-axis so that overlapping points can be seen

Analysis

Although the ANOVA for this model is numerically very similar to that of the fixed effects model, one difference is the estimation of 'variance components' associated with the within-mouse and between-mouse sources of variation. The one-way ANOVA table (Table 4.7) is used to estimate the 'variance components' from the mean squares for mice and error and the numbers of animals in each

Table 4.7 ANOVA of the data shown in Table 4.6

Source	DF	SS	MS	F	P
Mouse	4	1370.7	342.67	3.463	0.003
Error	15	1484.2	98.95		
Total	19	2854.9			
	Est. Var. Comp.				
Mouse	60.93 (38%)				
Error	98.95 (62%)				

group. Further details are given in some textbooks (e.g. Dixon & Massey 1983, Snedecor & Cochran 1980).

The between-mice component of variance is $(342.67-98.95)/4 = 60.93$.

The estimated within-mouse variance component is 98.95, the error mean square.

The total variance is $60.93 + 98.95 = 159.88$.

Thus $(60.93/159.88) = 38.1\%$ of the total variation is due to differences between mice, and 61.9% is due to variation within mice (between microscope fields). The F value to test the null hypothesis that there is no between-mice variation yields a *P* value of 0.003, suggesting that the mice differ in the density of their neurones, although most of the variation is due to differences among microscope fields. The hippocampus is a very heterogeneous structure and it may be better to measure more microscope fields per mouse, rather than use more mice. More details of these types of calculations are given by Snedecor & Cochran (1980; p 452). However, some account also needs to be taken of the costs. If mice are expensive (as they may be in terms of pain and suffering), there may be a case for making more determinations per mouse rather than using more mice. However, if determinations are expensive, then use of more mice may be appropriate.

Note that, as opposed to the fixed effects model, the mice are assumed to be a random sample of all mice of this type, and therefore give a direct estimate of the population variability of mice. However, in the fixed effect model, the conclusions of the analysis obviously cannot be extended with much confidence to other treatments not included in the experiment.

Fixed and random effects models are found in all the other experimental designs discussed below, though this may not be immediately obvious since it is not explicitly stated in the name of the design. For example, the randomized block design usually involves a specific treatment (a fixed effect) and a random effect (the block) in a single experiment.

The randomized block design

Blocking is an important and powerful tool for increasing precision at little or no extra cost. Yet the evidence is that it continues to be under-used in the biomedical literature (Festing 1994b). It is therefore important to understand how it works and when it should be used.

Blocking deals with known (or suspected) heterogeneity, but rather than simply spreading it out between treatments by random allocation, it is dealt with in a systematic fashion. Blocking improves precision by breaking up an experiment into a number of 'mini-experiments', each of which can be done on subsets of the experimental units which are more homogeneous than the whole. An example might be to remove the effect of rabbit size when comparing two treatments which may be size-related. The rabbits are divided into groups (blocks) of large and small rabbits, typically with equal numbers in each group, before the experiment is started. Usually between two and about six blocks could be used, depending on the range of body weights and number of rabbits. The number of animals in a block must not be less than the number of treatments, but having exactly one animal on each treatment group in each block is quite acceptable. Treatments are then compared within each block, with the results being combined to give the overall treatment means. The differences between blocks are eliminated in the statistical analysis. This will increase the power of the experiment, if the character being measured is in any way associated with the size of the rabbits, without the need to find more experimental units, i.e. rabbits.

Blocking deals with categorical factors such as 'large' and 'small', not with numerical variables. There are many situations where blocks could be used in biomedical and zoological research. The animals within a litter may be more similar than average, so it may make sense to treat them as blocks by doing within-litter experiments, assuming pedigree information is available. Time is often used as a blocking factor, because in a large experiment it is often impossible to treat or monitor all the animals at the same time, and responses as well as instruments and environments can drift over time. The supply of animals may also be a limiting factor. Mutant and transgenic animals may not breed well, so only small numbers of animals (enough for a block, but not for a whole experiment) are available at any one time.

People may also differ in the way that they carry out the treatment (e.g. training the animal for a behavioural task, injection, surgical procedure, etc.), handle the animals and record data (particularly behavioural data). Thus if more than one researcher is involved in the experiment, it may be worthwhile blocking on personnel.

Finally, practical considerations rather than a desire to increase precision may govern the use of blocks. A large study may require that the animals are observed in different counties (in the case of fieldwork), farms, or in different rooms of an animal house and may therefore be subject to different environmental conditions (e.g. weather, temperature settings). Animal house rooms may be looked after by different technicians. Cage position in the room may also be relevant, as the cages in the different rows are likely to experience different light levels and ambient temperatures. All these factors can be treated as blocks.

Statistical analysis of a randomized block design

A two-way ANOVA 'without interaction' is used for analysing a randomized block experimental design. Whereas treatments are a fixed effect, i.e. the levels are determined by the investigator, blocks are a random effect because the investigator cannot determine the mean level of each block. Thus in this design there is a source of variation associated with treatments, a source of variation associated with blocks, and the blocks × treatments interaction provides an estimate of the error variation, and is used for the error term.

The data in Table 4.8 come from a multi-laboratory study of mutations in transgenic mice. Control and two dose levels of a known mutagen were given to the transgenic mice, and the DNA was extracted and sent to five laboratories for a blind trial of inter-laboratory variability. Each laboratory estimated the number of mutations at a particular genetic locus in six samples of DNA, two

Table 4.8 Number of mutations found following treatment of transgenic mice with a known carcinogen when assessed in five different laboratories

Laboratory	Dose 1 (control)	Dose 2	Dose 3
1	11.8	18.8	16.4
1	11.3	20.6	23.4
2	10.3	16.8	13.2
2	9.5	13.0	14.6
3	6.5	2.4	8.4
3	3.9	3.9	8.6
4	5.4	11.7	12.4
4	6.2	11.6	12.2
5	14.2	21.2	22.5
5	15.2	13.7	15.8

from each dose level. In this case, laboratories are a random variable and are used as the blocking factor if the aim is to see whether there is any evidence for treatment differences in the number of mutations.

The raw data, not taking account of the identity of each laboratory, are shown graphically in Figure 4.5. A one-way analysis of variance of these data (not shown) would show that differences among treatments are not statistically significant ($F_{2,27} = 2.80$, $P = 0.078$). Figure 4.6 shows the same data, but now separate symbols are used for each laboratory. Note that within each laboratory there is some degree of consistency. Much of the variation is due to differences between laboratories. Figure 4.7 shows the data again, but adjusted for differences between laboratories. This inter-laboratory variation can be eliminated by the randomized block statistical analysis, shown in Table 4.9, using MINITAB. This shows first the ANOVA table, then the means for each dose, and finally Tukey's test to compare the individual means to see which differ. In this case, dose level one differs from levels two and three, but these do not differ from each other.

Differences amongst laboratories are of little interest, but the effect of removing this variation by blocking is immediate: the variance goes down from 27.1 with 27 DF (calculations not shown) to 6.85 with 23 DF. Note that the pooled variance is the error mean square in the ANOVA table. Without taking account of the blocks,

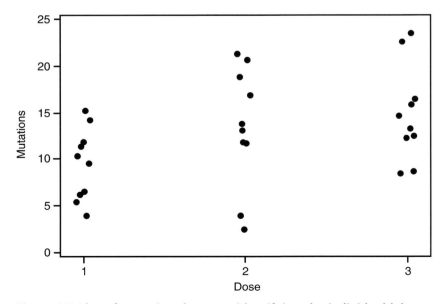

Figure 4.5 Plot of mutation data, not identifying the individual laboratories. On the X-axis are the three dose levels and on the Y-axis the number of mutations. Note the substantial variation within each dose group

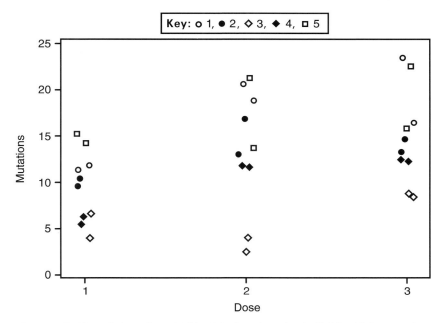

Figure 4.6 Plot of data from Table 4.8 showing the individual laboratories. Note that much of the variation within groups is due to differences between laboratories

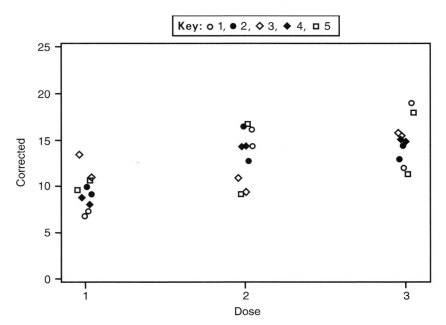

Figure 4.7 Randomized block design. Variation reduced by subtracting block means from each observation and adding in the overall mean

Table 4.9 ANOVA table for mutation numbers. Blocking for inter-laboratory variation

Analysis of variance for mutation numbers

Source	DF	SS	MS	F	P
Lab	4	576.45	144.11	21.03	0.000
Dose	2	152.43	76.22	11.12	0.000
Error	23	157.64	6.85		
Total	29	886.52			

Means for mutation numbers

Dose	Mean
1	9.430
2	13.370
3	14.750

Tukey Simultaneous Tests
Response Variable mutations
All Pairwise Comparisons among Levels of Dose

Dose = 1 subtracted from:

Level Dose	Difference of means	SE of difference	Adjusted T-value	P value
2	3.940	1.171	3.365	0.0073
3	5.320	1.171	4.544	0.0004

Dose = 2 subtracted from:

Level Dose	Difference of means	SE of difference	Adjusted T-value	P value
3	1.380	1.171	1.179	0.4774

differences among the treatments were not significant, but after taking account of the blocking the differences were highly significant. There is a cost to blocking: four degrees of freedom were spent in fitting the five laboratory means. But in this case, the cost of blocking was more than offset by a very substantial increase in the precision of the estimates of the treatment effects. Of course, if there were no inter-laboratory variation, blocking would not have been successful in reducing the residual sum of squares.

Assumptions

The usual assumptions of homogeneous variances and normality of the residuals can be checked using residuals plots, as already described. If the data do not appear to fit the assumptions, a transformation of the data (say taking logarithms or square roots) may help to normalize the data. Further details are given in most statistical texts. With these data there is no evidence of non-normality or heterogeneity of variance.

Missing observations can cause problems, although most modern statistical packages will do a 'least squares' analysis, which takes account of unequal numbers.

There is a further assumption with a randomized block design. It is that there is no block by treatment interaction, i.e. the response is the same in each block (apart from random variation). For example, this situation would arise if the mutagenesis treatment had no effect in some of the laboratories, but a large effect in others. Averaging across all laboratories may mean that the treatment effect is missed. If such an interaction is suspected a transformation of the data may correct it.

The Latin square and similar designs

The randomized block design described above removed one systematic source of variation (laboratory differences) in addition to the treatments. But suppose there are two or even three sources of potential heterogeneity such as laboratories, time at which the determinations were carried out, and technicians doing the work. Latin and Graeco-Latin squares are special cases of blocking designs where the aim is to control two (Latin square design) or three (Graeco-Latin square designs) simultaneous sources of variation. Graeco-Latin squares are rarely used in animal experiments.

When measuring learning ability in a maze there may be an effect due to time of day, caused by circadian rhythms, and there may also be some effect due to the variation in body weight of the mice. One limitation of the ordinary Latin square is that the number of experimental units should be the square of the number of treatment groups. Thus, if there are three treatments (no running, moderate running and marathon running), the Latin square should consist of nine animals (3×3). The time of day would be considered to be the rows of the experiment, and the body weights would be the columns. Three small animals (one from each of the three treatment groups) are measured, say, between 09.00 and 10.00 h, three middle-sized ones between 11.00 and 12.00 h, and three of the largest between 14.00 and 15.00 h. As mice are nocturnal animals, it might make sense to do these experiments using a reversed lighting schedule. However, with this experiment there will be two degrees of freedom associated with rows, two with columns, two with treatments, leaving only two degrees of freedom for error. Latin squares can be replicated to increase precision. In this example the 3×3 design could be done three times, giving six degrees of freedom for error. Normally, an un-replicated square should have between five and eight treatments: any less, and the experiment will lack precision; any more and it will become excessively large and difficult to manage. The statistical analysis of Latin squares presents no problems provided there are no missing values. A few can be tolerated by most computer programs, but a design of this sort should not be used where missing values are likely.

Latin squares are uncommon in experimental work with animals, but they could be a useful way of increasing precision and/or reducing animal numbers in some circumstances. For example, this design has been used in studying the effect of orbital sinus bleeding on some behavioural variables in mice (van Herck *et al.* 2000), with each of three mice receiving one of three different treatments at different times (so the mice were 'columns'), with the day of the week (Monday, Wednesday and Friday) being 'rows', and treatments being three different orbital sinus bleeding regimens. These designs are discussed in more detail by Cochran & Cox (1957).

Within-animal testing

The crossover design

In this design, two or more treatments are applied sequentially to the same subject. Thus, the experimental unit is the animal (or other subject) for a period of time. This design is closely related to a randomized block and Latin square, depending on exactly how it is designed. If treatments are applied in a random order, then the design is equivalent to a randomized block design where the animal is the block. The design is also sometimes used in situations where one treatment must always be given first because the other treatment permanently changes the animal. In this case there is an assumption that there is no time trend. However, if there may be a time trend within each animal (say as the animal ages or stops growing), a Latin square layout may be used. Suppose, for example, that there were three treatments applied to three animals in three periods. The animals represent the rows of the Latin square design, the time period the columns, and the treatments need to be arranged so that they are exactly balanced across periods (Figure 4.8).

The advantage of crossover designs is that differences between individuals are likely to be quite large, but variation within an individual may be quite small, so treatment differences are estimated with high precision. The disadvantage is that treatment effects may carry over from one period to another, so a 'washout' or rest period between treatments is usually required. This limits the applicability of these designs, as the effects of the treatments must be temporary and reversible. Hence these designs may be appro-priate for testing drug treatments, but not for certain surgical

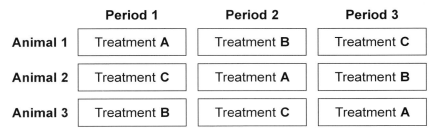

Figure 4.8 Diagram of a crossover design set out as a Latin square. Rows are represented by animals, columns by time. Note that rows and columns are balanced with respect to the treatments, i.e. each row and column has exactly one of each treatment. A crossover design can also be done as a randomized block if it is thought that there is no time trend. In this case treatments can be given in random order and the other restrictions of a Latin square will not apply

interventions. Crossover designs are used widely in clinical trials, which confirms their ethical validity under the right circumstances. They are also used in some pharmacological and physiological experiments with animals under anaesthesia without recovery. In these circumstances, several drugs can be given to the animals sequentially (with appropriate washout periods), with, for example, ECG or EEG responses being monitored for each drug. These designs are under-utilized in animal experiments, where they have the potential to reduce animal use considerably in situations where the variability between animals is large.

Balanced incomplete block design

Sometimes a natural block cannot accommodate all the treatment groups. Litter size might be a blocking factor, but this could be a problem if there are eight treatment groups but it is not possible to get litters of more than four animals. With a crossover design it may not be possible to test all the treatments within the same animal. This may be due to time restrictions, as perhaps it would take a long time to study a reasonable number of treatments in a full crossover study with appropriate rest periods being allowed for washout. Alternatively, there may be physical restrictions on the number of occasions or sites on which an animal may be used. A full description of incomplete block designs is beyond the scope of this book, but designs for various block sizes and number of treatments are given by Cochran & Cox (1957).

Repeated measurements designs

Two or more (often many more) measurements may be made on each animal over a period of time. For example, 20 rats might be assigned to two drug treatment groups (one of which is the control), and blood samples are taken to obtain a blood 'count'. If a lot of blood is needed, it may be necessary to kill some of the rats at each time point. The experiment would then be a 2 (drugs) $\times$ 2 (times) factorial design (discussed below). However, if only a small quantity of blood is needed, then each rat could be anaesthetized and bled at each time point. The experiment should then be analysed as a repeated measures design.

Unfortunately, repeated measurements are often made without full thought as to how they might be analysed. Time is an unusual factor as it cannot be randomized. If an animal is injected with a drug, it is not possible to study its response after one hour before studying its response after 10 minutes. This causes some conceptual problems with repeated measures designs. It is not entirely acceptable to perform a multiple series of analyses at each time point, as the data collected from an animal at a given time point will be correlated with its data at other time points, so the multiple analyses will also be correlated. Here are some more acceptable approaches:

Analyse a preselected time point only

The simplest approach is to analyse the data for one time point only, chosen before the start of the experiment. However, it is not always possible to use this approach if there is uncertainty about the likely time to onset of a treatment effect.

Find an informative summary measure

Some possible summary measurements are indicated in Figure 4.9. Profile (a) might be best summarized by calculating a mean response over time for each animal. Profile (b) might be summarized by estimating the slope (rate of change) for each animal. Profile (c) shows the area under the curve (AUC) used as a summary of the magnitude of the response for that animal. Profile (d) shows the time to peak response.

With the rat bleeding example, if the main aim of the experiment is to look at any changes in blood count, then it might be better to analyse the difference in blood count as a summary measure in a simple one-way ANOVA, or to analyse the final blood count correcting for the initial blood count by the analysis of covariance.

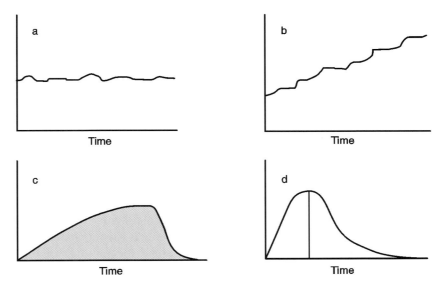

Figure 4.9 Some possible summary measures with a repeated measures design. In case (a) the mean of all measurements may be most appropriate. In case (b) the appropriate summary measure could be the slope of the line. In case (c) the area under the curve could be used, and in case (d) the time to peak response could be used

When there are more than two measurements per animal, it is possible to perform a multivariate analysis of variance (MANOVA) which regards the data at each time point as a separate character (dependent variable), but takes account of possible correlations between characters.

Repeated measures ANOVA

There is some confusion in the literature about repeated measures ANOVA. Some authors (e.g. Howell 1999) use the term to cover what was described in the previous section as a crossover design on the grounds that the same individual is measured more than once. However, more commonly, time is regarded as a factor in the ANOVA, so that there may be an analysis involving, say, animals on two or more treatments, measured at various times with a time × treatment interaction. The latter, if significant, would supposedly indicate that the response to some treatments changes with time differently from other treatments. However, as noted above, such an analysis is of debatable validity (Mead 1988, p 407), and expert advice is strongly recommended. A useful practical account of the analysis of repeated measures designs is given by Everitt (1995).

More complex designs

Factorial designs

Often the simultaneous effects of more than one factor such as a treatment and sex (both independent variables) on the response (dependent) variable need to be investigated. For example, it might be suspected that a new drug for reducing blood pressure only works in one sex, or it may be desirable to find out whether the response to a dietary treatment is the same in animals of different genotypes. The difference between a factorial design and a design involving blocks is twofold. First, blocks are random effects whereas a factorial design involves two or more fixed effects such as treatment and sex. Secondly, while the aim of blocking is to increase precision, the aim of the factorial design is to increase the amount of information and increase the 'generality' of the results. An experiment which explores whether a treatment effect is observed in both sexes has wider generality than one involving only one sex. The use of a factorial design can never lose information, even if there are no interactions, and when there are interactions the design is indispensable because it is the only way of showing that such interactions exist.

Factorial designs and randomized block designs are not mutually exclusive. Factorial experiments are often designed as randomized block designs. An example is given below.

Interactions

Biologists often use words such as synergism or potentiation to describe what statisticians call interactions. Thus, if a response to a drug is seen only in alcoholics it could be said that alcoholism potentiates the effects of the drug, or that there is an interaction between alcoholism and the drug treatment. Usually, factorial designs are used to show whether the response to a treatment is the same across all levels of whatever other factors are being studied. Several factors can be included. If the response to the above drug was only seen in male alcoholics, but not in females or male non-alcoholics, then that would be a three-way interaction of drug × drinking × gender. However, 'high order' interactions of this sort are relatively rare, their detection may require large experiments, and where found they are often difficult to interpret.

The ANOVA can handle complex factorial designs with several factors and the interactions between them, and other design considerations such as blocking. In some cases high order interactions will be of little interest and, as they are usually not statistically significant, it is not always necessary to estimate them.

Table 4.10   Learning ability scores in three strains of mice subjected to two running treatments

Strain	Learning score	
	No running	Running
A	8.55	13.59
A	9.63	15.05
A	10.33	15.14
A	12.04	16.83
B	8.43	10.69
B	10.10	11.32
B	10.64	11.41
B	11.33	11.76
C	7.22	9.86
C	8.21	10.52
C	9.91	10.67
C	10.75	12.84
Mean A	10.14	15.15
Mean B	10.13	11.30
Mean C	9.02	10.97

As an example of a completely randomized factorial design with two factors, suppose that the factors of interest are physical activity (inactive and active) and mouse strain in order to test whether (1) the running activity affects learning ability (as before), and (2) whether any response was strain-dependent. Three strains of mice, referred to here as strain A, B and C, were therefore tested. Eight mice were tested per strain, four of which were allowed to run on a wheel for a fixed amount of time, and four (the controls) were not (24 mice in total, *see* Table 4.10). Note that even without a statistical analysis there is a clear suggestion that strain A responds differently to running activity than the other two.

Analysis

The ANOVA is shown in Table 4.11. Plots of fits and residuals (not shown) showed no evidence of non-normality of residuals or

Table 4.11   ANOVA table for factorial design (SPSS output). Note that 'running: strains' is the interaction term sometimes designated 'running × strains'

Source	DF	SS	MS	F	P
Running	1	44.119	44.119	26.838	0.000
Strains	2	30.030	15.015	9.134	0.002
Running : strains	2	16.524	8.262	5.026	0.018
Error	18	29.590	1.644		
Total	23	120.263			

Pooled within-group standard deviation is 1.28, the square root of the error mean square heterogeneity of variances. Note that the strain × activity interaction effect is statistically significant at $P = 0.018$, suggesting that there is some evidence that strains differ in their response to the running activity treatment, as noted. *Post-hoc* comparisons could be used to explore these differences in more detail, but it might be sufficient just to know that there are slight, but statistically detectable strain differences in the response. Note that the pooled standard deviation is obtained as the square root of the error mean square.

The advantage of this factorial design over a single-factor design using only one strain is that it gives the additional information that response is, to some extent, strain-dependent, with an estimate of how much strains vary. The extra cost is negligible. The experiment used only 24 mice altogether, and about this number would probably have been needed had a single strain been used (but see Chapter 5 for a discussion of sample size estimation). If a single strain had been used, then the estimate of the increase in learning ability would only have been valid for that strain, and it would not have been known whether strains differed in response.

If there is an interaction between two ordered factors, it should appear as non-parallel lines (linear) or curves (if there are more than two treatments) when the means are plotted, as shown in Figure 4.10.

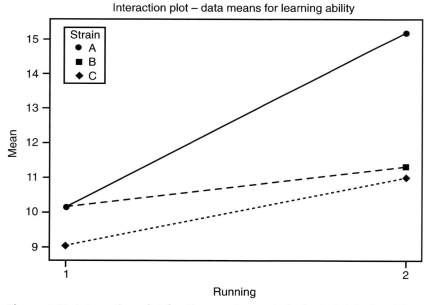

Figure 4.10 **Interaction plot for the running × strain factorial design.** The significant interaction effect implies that the three lines are not parallel, with strain A responding more than the other two strains to the running

Fractional factorial designs can be used to screen out the effects of many factors simultaneously within the same experiment. As the name suggests, this makes it possible to sacrifice some of the information gained from an experiment by only running a fraction (i.e. 1/2, 1/4, 1/8th. of the number of animals) that would be needed for a full factorial design. The information that is sacrificed is usually that about the higher order interactions of four or more factors, which are generally difficult to interpret anyway. A statistician or a good statistical textbook will provide help on these very efficient designs, and there are specialist statistical packages that provide step-by-step guidance on the use of full/fractional factorial designs, such as Design Expert, produced by Statease Corporation (*www.statease.com*).

Split-plot designs for combining between-animal and within-animal testing

Split-plot designs were originally developed by agricultural scientists doing field trials on crops. Trials (say of wheat varieties) would normally be done using a number of plots in a field, but for practical reasons the plots sometimes had to be quite large because machinery for sowing the wheat was large. However, fertilizer could be spread accurately on a much smaller area, so the plots with a particular variety of wheat could be subdivided or 'split'. Thus, this design actually involves two types of experimental unit. Whole plots are the experimental unit for comparing wheat varieties, but split-plots (e.g. half a plot) are the experimental units for comparing fertilizer treatments within a variety. The design has the great advantage that any variety × fertilizer interaction term is estimated quite accurately, so in that respect the design resembles a factorial design, and technically it is in fact a type of factorial design.

For a biological example, consider a behavioural experiment where animals are randomly assigned to treatment 1, an agonist that is believed to increase the number of mistakes that the trained animal makes, or treatment 2, a control treatment. So far this appears to be a completely randomized design. However, each animal also receives two short-term antagonists A and B, to decrease response, in a randomized order with appropriate wash out period between them.

The agonists treatments (1 & 2) are therefore compared between animals, while the effect of the antagonists (A & B) and how these interact with the different agonist treatments are compared within animals. The level of variability within an individual animal is often less than the variability between animals, so the analysis of variance table has to estimate both the between- and the within-animal variability, and test each treatment effect against the appropriate error term. Although such designs are quite useful, the statistical analysis is quite complex. A professional statistician should probably be consulted.

Sequential designs

With these designs, subjects or groups of subjects are tested at one time, in such a way that the result of the test is known before the next test is begun. The accumulated data are analysed, and a decision made regarding the next subject. This may involve alteration of dose (e.g. in the 'up-and-down' method of determining the LD_{50}) or, under some circumstances and designs, a decision either to terminate the experiment or to continue with another animal.

Sequential designs are appropriate only when test results can be obtained relatively rapidly, although they may also be well suited to experiments on large animals for which a non-sequential experiment would require excessive space. Both the 'up-and down' and the 'acute-toxic-class' (Schlede *et al.* 1992) methods for determination of the LD_{50} are sequential designs, and require smaller numbers of animals than the classical test (Lipnick *et al.* 1995). The relative merits of these designs should also take into account possible suffering when death is used as an end point. Sequential designs are optimum in some circumstances, and thus achieve the desired precision of results with smaller sample sizes than other designs. A sequential design to test the efficacy of compounds protecting against the effects of stroke in a rat model reduced animal use on average by about 35% (Waterton *et al.* 2000).

Construction and analysis of sequential designs is complex, and anyone who thinks such a design would be appropriate should seek professional advice. A monograph on sequential designs by Armitage (1975) is also available.

Reducing variability after the experiment: the analysis of covariance (ANCOVAR)

There are some situations where observations or measurements are made on the experimental subjects before the experiment is started, and the final observations need to be adjusted to take account of these observations. For example, when studying the effect of some treatment on the weight of an organ such as the liver, the analysis could be done on final liver weight, but it might be better to adjust for initial body weight because animals that are large to start with are likely to have a larger final liver weight. A common strategy is to divide the liver weights by the initial body weight. However, although this may work in some cases, in others it could actually be counterproductive. If there is, in fact, no association between initial body weight and final liver weight, then adjusting for body weight will actually decrease precision because dividing by an unrelated random variable will increase the variation in the final liver weight. Even if there is an association, dividing by body weight may not always be the most appropriate mathematical adjustment.

The analysis of covariance is designed specifically to deal with situations of this sort in an optimum way. What it does is first to determine whether there is a linear relationship between the covariate (e.g. the initial body weight) and the dependent variable (say liver weight) and, if so, it makes the optimum linear adjustment to the dependent variable, thereby reducing the error variation. The covariate should not itself be affected by the treatment.

As an example, the data in Table 4.12 come from an experiment that involved treatment with a compound believed to be a liver toxin. Thus, the hypothesis to be tested is that the compound does not affect liver weight. A one-way ANOVA shows that there are no significant differences ($F_{2,14} = 2.5$, $P = 0.15$) in liver weight between the three groups (group 3 being the controls). However, larger rats clearly have larger livers, though body weight itself is not affected by the treatments. An analysis of covariance was done using the MINITAB statistical package, with the results shown in Table 4.13. Now the covariate, body weight, is included in the first row of the ANOVA table, and its significance in association with liver weight is tested. In this case the association is significant at $P = 0.022$. The

Table 4.12 Body and liver weights (g) in three groups of rats. Treatment groups 1 and 2 have been treated with a potential liver toxin and group 3 is a control group

Treatment 1		Treatment 2		Treatment 3	
Body	Liver	Body	Liver	Body	Liver
238	8.97	219	11.11	229	8.40
262	9.62	230	10.22	276	10.51
239	10.34	219	10.19	250	10.65
271	11.14	251	12.57	276	9.12
229	10.49	245	11.80	247	9.65
		203	9.27		
		243	10.70		

Table 4.13 Analysis of covariance of the data given in Table 4.12. Note that MINITAB only does the analysis of covariance as part of the 'General linear model' ANOVA. This produces extra columns in the ANOVA table labelled 'Adj SS' etc, though in this case the adjusted SS (Adj SS) are identical with the SS. There is now one DF associated with body weight, and a P value of 0.022 showing that there is a significant association between body and liver weight.

Source	DF	SS	Adj SS	Adj MS	F	P
Body weight	1	0.4505	4.6932	4.6932	6.81	0.022
Treatment	2	8.4416	8.4416	4.2208	6.12	0.013
Error	13	8.9651	8.9651	0.6896		
Total	16	17.8573				

liver weights are then adjusted automatically to a common body weight, and the treatment differences are assessed and found to be statistically significant at $P = 0.013$. Thus, by adjusting the liver weight to take account of variation in body weight, a significant treatment effect is established. The next step would be to look at the individual means to assess the magnitude of the effect.

In some sense, the analysis of covariance can be considered more a method of analysis than a design since, as opposed to the blocking design, the adjustment for weight is done after the design has been specified and the experiment carried out. However, it is included here because, like blocking, it increases the precision of the experiment. It also forms part of the design in that a decision is normally taken before the experiment is done that covariance analysis will be used, and the covariate is measured accordingly. Covariance may have some advantages over blocking in that it uses the numerical data of the covariate, rather than the categorical data (small, medium, large) that would be used with blocking. However, it does not have some of the other advantages of blocking such as the ability to split the experiment into smaller parts which are more convenient to handle. Note that any covariate should not be subject to alteration by the treatment. In this example there is no evidence that body weight was affected by the treatments, though it might have been safer to use the body weight before the treatments were applied.

5

The determination of sample size

The determination of sample size is one of the most important, and difficult, decisions that a research worker has to make. If the sample size is unnecessarily large, then scientific resources will be wasted and animals will suffer unnecessary distress. However, if the sample size is too small, it may not be possible to achieve the scientific objectives of the study and there is a high chance that biologically important effects will not be detected. Unfortunately, there is no simple way of determining the optimum sample size, which depends on many factors, some of which are unknown at the time the experiment is planned. The aim of this chapter is to discuss some general principles which, if followed, should minimize the use of scientific resources (time, animals, cost of drugs, etc.) without compromising the scientific quality of the study.

Two methods of determining sample size, the 'power analysis' and the 'resource equation' methods are described. However, before introducing these methods, some factors that will influence sample size determination are discussed.

Factors affecting sample size

Objectives of the study

Studies may aim to estimate the mean of a particular character such as survival, ED_{50}, or tumour incidence; they may be designed as 'controlled' experiments to compare means of two or more treatment groups (some of which may be designated 'controls'); or they may be looking at the relationship between two variables such as the time course of response to a compound. Moreover, in some cases the aim will be to determine whether there are differences in means among treatment groups, whereas in other cases the aim may be to determine the magnitude of the differences, where it is obvious that they do differ. The optimum sample size will vary depending on these aims and on factors such as the number of different groups involved in the experiment. In this chapter it will be assumed that the experimental unit is an animal, but the same principles apply whatever is the actual unit.

There is a tendency to rely on what other people in the same discipline have done in the past. If most people have used a group size of six rats in similar studies in the past, the easiest course of action (and the one assumed to give the best possible chance of acceptance by a good journal) is to continue to use six rats per group. However, in many cases this will seriously overestimate, and in some cases underestimate the numbers of animals that should be used. Generally, the most important determinant of sample size is the number of animals available for the estimation of experimental error (i.e. inter-individual variation), and the required group size will decline as the number of treatment groups increases. In some circumstances, such as in regression studies, it is not uncommon to have a single observation at each time point or dose. This is widely accepted, provided there are enough time points altogether to give a good estimate of any regression relationships. A somewhat similar argument can be applied to controlled experiments where the more treatment groups there are, the fewer animals will be needed per treatment group. Research workers are strongly urged not just to follow previous research in their discipline, but to consider alternative approaches which may reduce the use of animals without reducing scientific output.

Type of data to be collected

Data can be either categorical, such as alive/dead, male/female, or they can be numerical. Numerical data can be discrete, such as counts (e.g. litter size or open-field activity), or continuous where something is measured and can take any intermediate value. In some cases, such as erythrocyte counts or even litter size in mice (which can range from about one to 15 or more), discrete data can be treated as being continuous.

Other types of data may also be collected. For example, individuals may be ranked, or the ratios of two objects may be expressed as a percentage. In some cases the data should be expressed as a score which may be assigned subjectively or with reference to some arbitrary scale. Data may also be 'censored'. For example, following some treatments the animals may be studied for a fixed period to see how long they take to develop tumours. However, no latency can be assigned to the animals which have not yet developed tumours. Similarly, litter size may be censored in that a litter size of zero does not really make sense.

The type of data largely determines the method of statistical analysis that will be used, and also the most appropriate sample size. Where a choice is possible, continuous numerical type data will usually result in smaller sample sizes than categorical data.

Some characters, such as the concentration of a blood enzyme, tend to have a skewed distribution, with most animals having low levels but some having extremely high levels. Such data cannot usually be analysed using parametric methods such as Student's *t-test* or the analysis of variance. They either require transformation to a different scale (probably by taking the logarithms of the enzyme determinations in the above example), the use of non-parametric statistical methods or special types of parametric test which have been 'fixed' to take account of heterogeneous data.

Uniformity of the experimental material

The more variable the experimental subjects, the larger the sample size needed to do a sensible experiment. This is largely because with heterogeneous material there is a real possibility that groups will differ to a material extent even before the experiment is started, so group size has to be increased to average out the differences. This was discussed in more detail in Chapters 3 and 4.

Design of the experiment

Formal experimental designs were discussed in Chapter 4. Choice of design depends mainly on the nature of the questions being asked and the biological situation, but the design will certainly affect the optimum sample size. In general, as the experiment becomes more complex, with more treatment groups, the number of animals per group can be reduced. This is particularly true with factorial designs, as noted above.

Determination of sample size using the power analysis or the resource equation methods

When the objective of the study has been clearly formulated, an appropriate experimental unit has been chosen, the type of data to be collected has been decided, the problem of obtaining uniform experimental units has been considered, the number and type of treatments has been decided, and a formal design chosen, it becomes possible to estimate the required sample size. This can be done using either the power analysis or the resource equation method.

Generally, a power analysis is appropriate for relatively simple but expensive experiments such as clinical trials or animal experiments which are likely to be repeated several times with slightly different treatments. The resource equation method is suitable for one-off complex biological experiments involving several treatment groups, quantitative data, possibly with many dependent variables, and/or where there is no estimate of the standard deviation.

Power analysis for determining sample size

The calculations for the power analysis can be complicated for any experiment more complex than a comparison between two samples using Student's *t-test*. Cohen (1988) provides extensive tables and a good explanation of power analysis methods. However, there are now some statistical computer packages such as MINITAB and SAS (SAS Worldwide Headquarters, SAS Campus Drive, Cary, NC 27513-2414, USA, *www.sas.com*) which will do power analysis/ sample size calculations for some designs. There are also some stand-alone programs such as nQuery Advisor (Statistical Solutions Ltd, 8 South Bank, Cross's Green, Cork, Ireland, tel $+$ 353 21 319629, *www.statsol.ie*), PASS 2000 (NCSS, 329 North 1000 East, Kaysville, Utah 84037, USA, *www.ncss.com*), and Power and Precision (Biostat, 14 North Dean Street, Englewood, NJ 07631, USA, *www.Power-Analysis.com*). A free evaluation copy can be downloaded from most of the Web sites. At the time of going to press a comprehensive list of commercial and free power analysis tools can be found on *www.forestry.ubc.ca/conservation/ power/index.html#samplepower*.

The power analysis method depends on the relationship between six variables:

(1)   the effect size of biological interest;
(2)   the standard deviation;
(3)   the significance level;
(4)   the desired power of the experiment;
(5)   the sample size;
(6)   the alternative hypothesis (i.e. a one- or two-sided test).

Fix any five of these, and a mathematical relationship can be used to estimate the sixth. Where an experiment involves measuring several different dependent variables such as body weight, haematological parameters and blood pressure, it is necessary to decide which of these characters is of most interest as this should be used to determine the sample size.

Effect size of biological interest

The first step is to decide how large a biological effect would be of scientific interest. The larger the effect size, the smaller the experiment will need to be to detect it.

For continuous characters it is often helpful to consider the percentage change in the mean that may be of interest. Would a 10% change in body weight in rats be of biological significance, and should the experiment be designed to detect it, or would it only be of importance if there were a 20% change? For categorical variables the difference in percent responders between groups that is of biological interest will need to be specified. If 50% of the control group is expected to show some effect, what proportion in a treated group would it be of interest to detect? Would it be 55%, 60% or 70%?

Although it may be difficult to decide how large a change is of interest, it must be done in order to use the power analysis method.

The standard deviation

The estimated sample size is heavily dependent on the size of the standard deviation. For discrete characters such as dead/alive, the standard deviation is a function of the proportion that die, so there is no need to specify it. For continuous characters such as body weight or enzyme concentration, it is necessary to have an estimate of the standard deviation. However, the standard deviation cannot be known exactly until the experiment has been completed. In many cases, similar experiments or reports of similar experiments in the literature can give an estimate of the standard deviation. If these are not available, then a pilot study may be necessary to provide an approximate estimate, although if the pilot study is small, the estimate may be inaccurate. It is generally worthwhile doing a 'best case' and a 'worst case' calculation based on the lowest and highest of the available estimates to see the effect of this on sample size estimates.

The significance level

The significance level is the probability that the experiment will give a false-positive result, i.e. that a 'statistically significant' effect is found when in fact it is entirely due to chance sampling error. This is often referred to as a Type I error. Commonly, a significance level of $P = 0.05$ is used, though there is nothing magic about this number. A significance level of 0.01 would imply that there is only one chance in one hundred of a false-positive. However, other things being equal, specifying a low chance of a false-positive result will increase the chance of a false-negative result (failing to detect a true biological effect, often referred to as a Type II error).

The power of the experiment

The power of the experiment is the probability of detecting the specified effect at the specified significance level. The general aim should be to have powerful experiments that have a high chance of detecting an effect if it exists. Somewhat arbitrarily, the power is usually set somewhere between 80% and 90%. The higher the power, the larger the sample size that will be needed, so specifying a very high power, such as 99%, may require an enormous number of animals.

Sample size

In many cases it is the sample size that is to be determined, and all the other variables are specified. However, sometimes the sample size is fixed by the number of animals that are available, in which case the power of the proposed experiment for the fixed sample size can be estimated. In some cases this power will be so low that the experiment may not be worth doing.

Determination of power is also useful in the interpretation of the results of a completed experiment that has produced a negative result. This may be real or it may have resulted from insufficient power, possibly due to too small a sample size. It may, for example, be possible to say that an experiment had a 90% chance of detecting a specified treatment effect, and as it did not, it is likely that an effect of this magnitude does not exist. On the other hand, it may be that an experiment had only a 40% chance of detecting the specified treatment effect, in which case such an effect might well exist.

The alternative hypothesis

The usual null hypothesis is that there are no differences among treatment means, with the alternative being that there are. This leads to a two-sided significance test. However, if the alternative is that means differ in a particular direction, then that will lead to a one-sided test.

Putting it together

As already noted, power analysis software is available from several sources. nQuery Advisor is one such program. It is a Windows program dedicated to power analysis calculations. The initial screen specifies the nature of the planned experiment. The type of data (means, proportions, survival, agreement, regression), the number of groups (one, two, more than two), and the type of statistical analysis (a significance test, a confidence interval or a test of equivalence) need to be specified. When comparing two means, Student's *t-test* or a non-parametric Wilcoxon/Mann-Whitney rank sum test can be specified. With more than two means, a one- or two-way analysis of variance will be assumed. When comparing two proportions, calculations based on a chi-squared test with or without a continuity correction can be specified.

After the method of analysis is specified, a screen is used to enter the variables. An example for a two-sample *t-test* is shown in Table 5.1. Note that the screen has several columns so that various alternative specifications can be compared. If, for example, the sample size as originally specified is impracticably large, it is possible to use some of the other columns to see how large the experiment would be if the standard deviation could be reduced by using more uniform animals, or if the effect size was specified to be larger than originally planned.

An example

Mice vary genetically in sleeping time under hexobarbital anaesthetic (Jay 1955). A study of five inbred strains and two outbred stocks, using 29 to 63 mice per strain, found a mean sleeping time ranging from 18 to 48 minutes in strains SWR and A/N, respectively (*see* Table 2.3 in Chapter 2, page 25). The standard deviation in sleeping time averaged 3.5 minutes in the five inbred strains and 13.5 minutes in the two outbred stocks. Inbred BALB/c mice and the outbred Swiss mice each slept for just over 40 minutes, but the standard deviation in the BALB/c mice was only two minutes, whereas it was 15 minutes in the Swiss mice.

Suppose the aim was to set up an experiment to see whether soft wood bedding material altered sleeping time in mice. How many mice would be needed in a two-group (hardwood versus softwood) comparison experiment?

Using nQuery Advisor, the data might be entered as shown in Table 5.1. The first column assumes that Swiss mice will be used. With:

Table 5.1   Printout of an nQuery Advisor screen used to estimate sample size for a two-sample *t-test*. For details see text

	Col. 1	Col. 2	Col. 3	Col. 4
Test significance level, α	0.050	0.050	0.050	0.050
1 or 2 sided test?	2	2	2	2
Group 1 mean, $\mu 1$	40.000	40.000	40.000	40.000
Group 2 mean, $\mu 2$	36.000	36.000	36.000	36.000
Difference in means, $\mu 1 - \mu 2$	4.000	4.000	4.000	4.000
Common standard deviation, σ	13.500	3.500	13.500	3.500
Effect size, $= |\mu 1 - \mu 2|/\sigma$	0.296	1.143	0.296	1.143
Power (%)	90	90	14	94
n per group	241	18	20	20

- a 5% significance level;
- a two-sided test (it is not known whether softwood will lengthen or shorten sleeping time);
- the mice on hard wood will sleep about 40 minutes, and those on softwood about 36 (or 44) minutes, a 10% change;
- the standard deviation is 13.5 minutes, the mean of the two outbred stocks;
- a power of 90% is specified.

As can be seen from Table 5.1, this means that the experiment would require 241 mice in each group, a very substantial and expensive experiment.

The numbers could be reduced by altering the specification. The power could be reduced to 80%, and/or the effect size of interest could be increased. However, a better alternative might be to reduce the variation among animals by using, say, the BALB/c inbred strain. This is shown in the second column. Note that the standard deviation used is the mean across all five inbred strains, rather than the two minutes actually found with this strain. Now this experiment could be done with only 18 mice per group, which emphasizes the importance of controlling variability using inbred strains, where this is possible.

Suppose it was decided to do the experiment with 20 mice per group, it is then possible to see what power the two experiments would have. This is shown in columns three and four. With the Swiss stock there would be just a 14% chance that the experiment could detect an effect of this size (assuming the rest of the specification stays the same), whereas with the BALB/c mice there would be a 94% chance of detecting the effect.

When an experiment has been done and no significant treatment effects are observed, the results may still be of interest. Non-significance could be because there really is no treatment effect, or it could be because the experiment is too small, for a given level of variability in the material, to detect an effect. A power analysis can be used to find out what effect size the experiment was capable of detecting by using a program such as nQuery Advisor and specifying the sample size used and the standard deviation actually found. A power and significance level can be chosen, and the program will then indicate the effect size. Alternatively the effect size that might be of interest could be specified, and the resulting power estimated. In this way negative results can sometimes be turned into positive statements.

As already noted, the power analysis method cannot be used if there is no estimate of the standard deviation for continuous characters. While in some cases this can be obtained from previous experiments or the literature, in others there is no such information. It is also difficult to use for complex experiments with several treatment groups because of the difficulty of specifying the effect size of biological interest. In these circumstances, the resource equation method, discussed below, can be used.

The resource equation method

The resource equation method can be used when the power analysis method is not possible, or practical or appropriate. Its main features are:

(1) It is easy to use for complex experiments where a power analysis would be difficult.
(2) It is only appropriate for experiments producing quantitative data that can be analysed by parametric methods such as the analysis of variance or the *t-test*.
(3) The method depends on the law of diminishing returns: increasing the size of an experiment beyond a certain point gives little extra information.
(4) It does not require an estimate of the standard deviation or the effect size of biological interest.
(5) However, with this method the power, significance level and alternative hypothesis are not specified.

In general, it probably gives smaller estimates of required sample size than the power analysis method, so is most appropriate where relatively large effects are likely, as is often the case with biological rather than clinical experiments. The method is used most appropriately when the aim of the experiment is to test an hypothesis, e.g. that there are no differences among treatment groups, rather than to estimate some parameter such as the size of the treatment effect.

Having used the method for an experiment, it remains possible to use a power analysis retrospectively to quantify the probability that the experiment could have detected a specified effect, as noted in the previous section.

Mead's (1988) 'Resource equation' is:

$$E = N - T - B$$

where E is the error degrees of freedom, N is the total degrees of freedom (i.e. the total number of experimental units minus one), T is the treatments degrees of freedom (the number of treatment combinations minus one), and B is the blocks degrees of freedom (the number of blocks minus one).

For experiments *not* using blocking:
$E =$ (total number of experimental units) $-$ (number of treatment combinations). Thus for an experiment with five treatment groups and six rats per group, $E = 30 - 5$.

Mead (1988) suggests that, as a general rule, E should be between 10 and 20. If E is less than 10, increasing the numbers would lead to good returns. If it is substantially more than 20, resources will be wasted. This can be seen from the shape of the curve of the 5% critical value of Student's t (Figure 5.1). Increasing the error degrees of freedom (E) from one to 10 leads to a substantial reduction in the

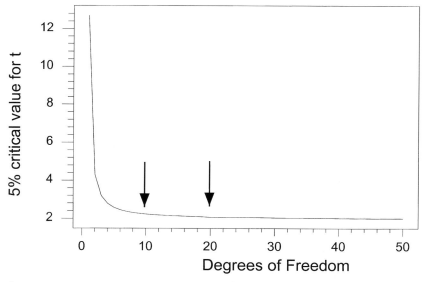

Figure 5.1 Five percent critical value of Student's t plotted against degrees of freedom. Arrows indicate points corresponding to 10 and 20 degrees of freedom. Note that after about 20 degrees of freedom the value of t needed for statistical significance hardly decreases

level of t that would be declared statistically significant. However, increasing E much beyond 20 hardly alters the critical value of t.

This is a rule-of-thumb method, and there are circumstances when it is justifiable for E to be larger than 20.

(1) When experimental units are cheap and non-sentient, such as wells in a culture dish or lower animals such as nematodes, then the cost (both financial and in terms of animal suffering) of a large experiment is negligible.

(2) When the experiment is quite complex involving several factors such as a treatment, both sexes, different time points, more than one genotype, etc. In these circumstances it is often desirable for the experiment to be balanced, with the same numbers in each treatment subgroup, and frequently this is not possible if E is fixed rigidly at 20 or less. In such circumstances it may be acceptable for E to range up to 40 or more.

(3) When the aim of the experiment is an accurate estimation of some parameter such as the magnitude of a treatment effect, rather than the testing of an hypothesis (that the treatment effect exists). Also, if small biological effects really are of interest it may be necessary to increase the sample size. Genetic linkage studies are a good example, where the interest is not so much whether two loci are linked, but how closely they are linked. This may involve large numbers of animals if the linkage estimate is to be really accurate. Where possible, the size of such experiments should be determined by a power analysis, rather than using the resource equation method.

An example

As an example, suppose it was planned to do an experiment involving four dose levels of a compound to determine its effect on 10 haematological parameters in rats. It is further decided that the experiment should involve both sexes, and two inbred rat strains in a factorial design in order to see whether the results are valid with both sexes and two strains. This experiment therefore involves $4(\text{doses}) \times 2(\text{sexes}) \times 2(\text{strains}) = 16$ treatment combinations with 10 dependent variables. It would be difficult to design such an experiment using a power analysis.

Using the resource equation method, with two rats per group $E = 32 - 16 = 16$. With three rats per group, $E = 48 - 16 = 32$. Thus, this method suggests that the experiment would probably be of an adequate size with a total of 32 rats. Note that many people would be horrified at the thought of doing an experiment with only two animals per group. However, this is a factorial design which typically requires much smaller numbers of animals per subgroup than single factor designs. In this experiment the means of the four dose levels of the compound (the estimates of most interest) would actually be based on eight rats, which is in reasonable agreement with numbers commonly used in non-factorial experiments. Note that had this experiment been designed simply to compare the four dose levels in a single sex and strain, then this method would suggest five or six rats per group. With five rats, $E = 20 - 4 = 16$.

There may be problems with the above experiment because it could be difficult to bleed all 32 rats within a short period, and blood parameters may alter over a period of time due to circadian rhythms. This would reduce the precision of the experiment. A randomized block design (see Chapter 4) could alleviate this problem. Sixteen of the rats (one from each group) could be bled first, say in the morning, with the other 16 being bled in the afternoon. In this case $E = 31 - 15 - 1 = 15$, so it would appear that using a randomized block design would result in E being slightly lower than if a completely randomized design had been used. While this is true, the increase in precision from using a randomized block design will normally more than compensate for the loss of a few degrees of freedom for error (E).

Note that with the sleeping time example discussed earlier, the estimate of group size using the power analysis method with an effect size of 15% (not the 10% shown in Table 5.1) varied from nine mice per group using the BALB/c mice to 108 mice per group with the Swiss mice (data not shown). Using the resource equation method, the estimate of the sample size under the same conditions should result in E being about 20 (maximum), and this will be obtained with 22 mice in total ($E = $ [total number of mice] − [number of treatment combinations]) or 11 mice per group. With this method, no account would be taken of the differences in variability between the strains. Assuming that isogenic mice are used, it gives an estimate that is approximately the same as the power analysis method for an effect size of 15%, though the power analysis can be adjusted more easily to any known conditions.

Conclusion on sample sizes

Determination of the most appropriate size for an experiment is not easy, but there are likely to be important benefits from keeping the experiment as small as possible, consistent with achieving the scientific objectives of the study. Excessively large experiments are not only ethically undesirable, but they also waste scientific resources. If two experiments could be done instead of one, using the same number of animals, then scientific output would increase, which would, in the long run, save animals and scientific resources. A general recommendation is to use the power analysis method if possible, otherwise use the resource equation method. In some circumstances, both methods can be used, with the final choice of sample size being a compromise between them, with some account being taken of previous experience.

6

Presenting and interpreting the results and making decisions

Introduction

It is not possible to go into great detail on how data should be presented. It will depend to a large extent on the purpose of the study, the nature of the results and the journal to which the paper is being submitted. However, given that the experiment may have caused pain and suffering in animals, it is essential that the data are well presented and that the conclusions are fully supported by the data. Otherwise the experiment may need to be repeated.

Description of the animals

The animals and the environment in which they were maintained should be fully described. Species, strain (using correct, internationally agreed nomenclature, where available; see Chapter 2), age, weight, sex, health status, caging, bedding, diet, temperature, humidity and light cycles should be noted. The 'Notes to authors' of a current issue of the journal *Laboratory Animals* can be referred to for more details of what is currently considered appropriate. Full details should be given of any surgical manipulations including methods used to collect body fluids. Experimental treatments should be described, including any anaesthesia and analgesia and methods used for euthanasia. The use of humane endpoints should be noted.

Data collection and statistical analysis

Small volumes of data will normally be recorded during the course of the experiment by hand in a book or on data sheets as specified in the experimental protocols. Annotations should include loss of animals or unexpected occurrences. Larger volumes of data are often generated automatically by machines, scanners, etc., and may be in the form of hard output or electronically stored files on disk or other medium. Where these data result from analysis of, for example, blood or other samples, the relevant links and identification with individual animals will also form part of the data record. These original data, including records of treatment and animal coding, provide the definitive record of the experiment and should be appropriately stored and not subject to alteration.

The data will need to be reproduced in an electronic format prior to statistical analysis. A spreadsheet may be used, prior to transfer to a dedicated statistical package, or the data can be entered directly into the statistical package.

Data should be screened for obvious errors before any statistical analysis is started. The most appropriate method depends on the nature of the data, but the main aim is to detect any serious outliers. These may be due to transcription or typographical errors. Plots of individual observations versus group means, or plots of one dependent variable versus another are helpful. A preliminary analysis of variance could be done, and the residual plots studied. These will often show up any outliers and departures from normality. The data file may be printed and compared with the original data to ensure that it has been accurately transcribed. Where any original observations are altered or omitted from the analysis, the reasons should be clearly indicated with cross-reference to the original data if necessary. This should be done in such a way that the statistical analysis can be done with and without the alterations.

All statistical analyses done on the computer should be printed out and annotated with the date on which the analysis was done and the name of the person who did it. The whole analysis should be kept with the experimental protocols and raw data so that a re-analysis can be done in the future should this be necessary.

Presentation of the results

The aim in presenting the results is to show the data as clearly and accurately as possible. Wherever possible, raw data should be given in tables (if there is only a little of it) or in charts and scatter diagrams. As a matter of principle, data should be manipulated as little as possible prior to the statistical analysis. 'Normalization' and expressing some group means as a percentage of the controls, for example, should be avoided if possible. Means and proportions should usually be presented with some measure of variation such as standard deviation, standard error, or confidence interval. Altman *et al.* (2000) suggest that, in order to avoid confusion, the $\pm$ symbol should be abandoned and means should be presented as '9.6 (SD 2.1) units'.

Where possible raw data should be shown in preference to error bars. Figures 6.1 and 6.2 show two plots of the same data. Figure 6.1 is preferable because it gives a truer impression of the real spread of the data points than Figure 6.2.

Standard deviations (SD) provide an indication of the variation between individual experimental units. They are independent of the numbers in the group, and are most appropriate if the aim is to indicate the variation in each group. However, the number of observations should also be stated. Provided residuals plots have shown that the variation is about the same in each group, the best estimate of the standard deviation is the square root of the error MS in the ANOVA table.

With a randomized block design the estimate of the standard deviation has to come from the square root of the error mean square in the ANOVA. With all designs, residuals plots (Chapter 4) should have been used to indicate whether the variation is approximately the same in each group. If it is, then again the best estimate of the standard deviation is the pooled estimate obtained from the error

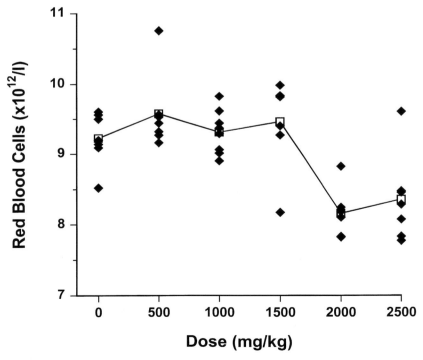

Figure 6.1 Red blood cell counts in mice following treatment with a potentially toxic compound at various dose levels. The plot shows the scores of individual mice with a line connecting means at each dose level

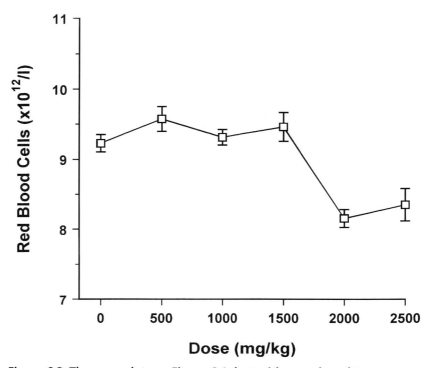

Figure 6.2 The same data as Figure 6.1, but with error bars ($\pm$ one standard error) around each mean. Figure 6.1 is preferable as it gives a more accurate representation of the actual data. This plot obscures the inter-individual variation, which may be of scientific interest

mean square in the ANOVA, and there seems to be little point in quoting a different SD for each mean. However, if the standard deviations differ among groups, this should be clearly indicated.

Standard errors of the mean (SEs or 'SEM'; it is better to add the 'M' to make it quite clear that these are not SDs) are the standard deviations of group means. Their magnitude depends on the standard deviation and 'N', the numbers in each group. People sometimes use them because they are smaller than the standard deviation, and therefore 'look better'. This argument is quite inappropriate. If SEMs are used, N must be stated for each mean. However, somebody wanting to judge whether two means differ significantly using the SEMs will have to do a *t-test*, so really the SEMs alone are not all that helpful, particularly if N differs much between groups. If means have already been compared as part of the statistical analysis, there seems to be little point in using SEMs. It would be best to indicate in a table whether the means are significantly different at, say, the 5% level of probability.

Confidence intervals (CI) provide the best indication of the precision of a mean or the difference between two means, taking into account the numbers in the group. Presenting, say, a 95% confidence interval saves the reader the trouble of doing a *t-test* either explicitly, or in his or her head. For this reason, the confidence interval is a much better indication of the precision of a mean or a difference between means than is the standard deviation or standard error.

Least squares means may be produced by two-way ANOVAs with unequal numbers in each group. These take account of the fact that means based on larger numbers are more accurate than those based on small numbers. They should be used when groups sizes are unequal.

Proportions and percentages, like means, should also be presented with some estimate of their reliability such as a standard error or, preferably, a confidence interval, or showing the results of a statistical test to indicate which are significantly different. Formulae are given in most statistics textbooks (Snedecor & Cochran 1980, Maxwell & Delaney 1989, Altman 1991).

Exact P values should be given in preference to $P < 0.05$ or the use of asterisks. This avoids claims that an effect is 'not significant' when $P = 0.056$. Remember that biological and statistical significance are not the same thing.

Contrasts, i.e. differences between means that are of particular interest, should have been decided at the design stage, and the statistical analysis should aim primarily to study these rather than using *post-hoc* comparisons of all means. If a dose-response is being studied, then an appropriate regression analysis should be performed rather than comparing each dose group with the controls. So-called 'orthogonal contrasts' (Chapter 4, Snedecor & Cochran 1980, Altman 1991) can often be used in the ANOVA to see whether there is a significant linear trend across dose groups, and whether there is significant non-linearity in the response. However, such analyses are not normally available except in advanced computer packages, so the advice of a professional statistician should be sought.

Exploratory versus confirmatory studies

An experiment is often designed both to test some clearly defined hypothesis stated at the time the experiment is being designed (i.e. it is a confirmatory experiment) and to explore relationships among variables after the experiment has been done (i.e. as an exploratory experiment). Exploratory data analysis of large sets of data can throw up apparently 'significant' effects just by chance, and it is often difficult to take account of such effects by adjusting significance levels using, for example, a Bonferroni correction (Maxwell & Delaney 1989). Exploratory data analysis is perfectly permissible, and useful, provided it is clearly understood that its main aim is to provide hypotheses which can be studied in future experiments, rather than to provide definitive tests of hypotheses from the present experiment.

People sometimes design experiments with, say, six rats per group because a SD or SEM based on fewer numbers is less reliable, and they may be criticized by referees if they use fewer animals per group. This should be firmly resisted. Where possible, sample sizes should be determined using either the power analysis or the resource equation methods, as outlined in this book, and pooled estimates of the standard deviation should normally be used in any case.

Transformed data

Data sometimes need to be transformed before they can be analysed using a parametric method such as the *t-test* or the ANOVA. Concentrations of substances in body fluids, for example, often have a log-normal distribution which can be normalized by taking the $\log_{10}$ or $\log_e$ of the individual observations. An angular transformation (Snedecor & Cochran 1980) may need to be used for percentage data where many of the values are less than 20% or more than 80%, and a square root transformation is appropriate for counts, when the mean count is relatively low and the resulting data have a Poisson distribution. However, people are not used to these scales, and may object to means being shown on a log or square root scale. In these circumstances the analysis should be done on the transformed scale, but the results will need to be presented on the original scale with an indication of statistical significance being given. In some cases this will involve a transformation of means back to the original scale, say using antilogs, if a log transformation has been used.

Many dependent variables

Some experiments involve more than one dependent variable. For example, an experiment involving measurement of haematology and blood biochemistry may have 10–20 or more dependent variables such as red blood cells, haematocrit, lymphocytes, etc. Modern studies using gene arrays to assess changes in mRNA may involve several thousand dependent variables measured on each experimental unit. This does not obviate the need for the several types of designs discussed here. Such studies often require special multivariate statistical analysis such as 'principal components analysis', 'discriminant function analysis', and various clustering methods which take account of any relationships between the variables, and can reduce large quantities of raw data down to a level where the results can be interpreted. These are specialized methods which require statistical advice.

Tables and figures

Tables and figures should be designed to convey information as clearly as possible. Computers often spew out numbers with many decimal places. However, in most cases two or three significant digits (i.e. the left-hand digits in a number) will suffice. It is easier to compare a set of means when they are in the same column, rather than in the same row. When means are shown in columns, standard deviations should be in a different column, and not in the same column but on a different row. Bar diagrams often take up a lot of space compared with a table of just a few means, so should be used sparingly.

Making decisions

Decisions have to be made at many stages in the design, execution, and interpretation of experimental studies. Many of these, such as the number of treatment groups, are usually made on a subjective basis depending on the experience and training of the investigator. In some cases, important decisions that can affect the outcome of the study are made by regulatory authorities, or are imbedded in legislation designed to protect experimental animals. In some cases, decisions also need to be taken on whether to continue with the experiment. Clinical trials are usually done 'blind', but often somebody looks at the unblinded results at an intermediate stage just in case there are large differences between the treatment groups, in which case the experiment may be terminated. With sequential designs, decisions on whether to continue are built into the design of the experiment.

However, in many cases an experiment is set up with the express purpose of reaching a decision based on the results. Thus, the experiment may indicate the relative potency of a biological preparation such as an antigen or vaccine, it may be designed to detect whether or not a chemical is carcinogenic, or it may involve a comparison of two surgical procedures, the aim being to choose one for future work. In interpreting such experiments, it is important to understand the sorts of statistical errors that can occur, and to take account of the possible costs associated with them. The term 'statistical errors' in this case does not mean errors arising from badly designed or analysed experiments, but the sorts of errors that can arise even when the experiment is well designed and analysed. These are discussed below.

Mistakes that may be associated with choice of fixed effects

In every experiment decisions will have been made about a large number of variables that may have influenced the outcome of the experiment. As an example, consider a single-factor experiment such as a study of whether or not chemical X is a carcinogen. This will involve choosing animals of a particular species, strain, sex and age or weight. The animals will be housed in groups or singly in cages with a particular bedding material, and they will be fed a specified diet. At the end of the experiment, the incidence of cancer in the control group will be compared with that of the treated groups, using an appropriate statistical test, from which a conclusion might be reached as to whether the chemical appears to be a carcinogen. A confidence limit on the results may even be estimated. In some cases highly sophisticated statistical methods are used in an attempt to extrapolate to a dose level which it is thought would produce virtually no cancer in humans.

However, no statistical test can indicate whether the same conclusion might have been reached had a different species, strain, sex, diet, bedding material, or grouping of animals been used. Any one of these variables could dramatically alter the outcome of the experiment. A more sensitive strain may well give a positive result with a weak carcinogen, an insensitive strain may fail to detect even a moderately strong carcinogen, and so on for many other variables. It makes little sense to use sophisticated statistical methods to extrapolate to a no-effect level when the effects of so many fixed factors are unexplored and unknown.

Strictly, the results of an experiment done with male F344 rats housed singly on softwood bedding are only predictive for other male F344 rats housed in the same way. However, science would make no progress if experiments were interpreted so strictly. The response of rats to a potential carcinogen is one of the main tools for assessing human risk. What scientists, and statisticians, should do is give a bit more thought to this problem. The effects of some of the variables that may affect the outcome of their experiments could be explored using factorial experimental designs, as discussed in Chapter 4. This can usually be done without using any more animals, and it may give much more useful and reliable results. If one of the factors does *not* influence the outcome, then nothing has been lost from including it in the study. The groups can be pooled, and the sample size is virtually unchanged. However, if one of the factors does influence the results, then this needs to be taken into account in the final interpretation.

The costs of making the wrong decision must also be considered. If a human carcinogen is not detected because the scientists used an insensitive strain of rats, then there will be an associated cost to society, individual humans and, possibly, the manufacturer of the chemical.

Mistakes associated with random effects: statistical errors

Even with well designed experiments, mistakes can occur in the interpretation of individual experiments due to sampling errors. Suppose, for example, that an experiment using rats is set up to determine whether compound X is carcinogenic, with a control and a treated group (for simplicity). The null hypothesis (H0) is that there is no difference among treatment groups, with the alternative hypothesis that there is a difference (a two-sided test because the compound may protect against cancer). The true 'state of nature', and the conclusions that might be drawn from the experiment are shown in Table 6.1.

Table 6.1   Statistical errors in hypothesis testing

State of nature	Conclusion	
	Reject H0	Accept H0
H0 is true	Type I error	Correct decision
H0 is false	Correct decision	Type II error

If the compound really is a carcinogen, and the experiment shows this, or if it really is a non-carcinogen, and the experiment shows this, then the correct conclusion will have been drawn. The chance that the experiment will correctly identify a carcinogen is designated the *sensitivity* of the test, and the chance that it will correctly give a negative result with a non-carcinogen is the *specificity* of the test. However, if it is a non-carcinogen but, due to unfortunate sampling variation, there are significantly more animals with cancer in the treated group, then the result will be a false-positive conclusion or a 'Type I' error. Similarly, a false-negative conclusion, or Type II error will be drawn if the experiment fails to detect a true carcinogen, due to chance sampling effects.

In reaching a decision on what to do about the chemical, it is clearly important to know the probability of both Type I and Type II errors. As a first approximation, the Type I error rate is determined by the significance level, so is usually not a problem. If the results are significant at the 5% level of probability, then that implies a 5% chance that this is a false-positive result and a 95% chance that it is not. The Type II error rate is 1−(the power), where power is as defined in Chapter 5. So, if the experiment finds no significant increase in the number of tumours, it is possible to do a power analysis to determine the power that the experiment had to detect a certain specified increase in tumour numbers, given the size of the experiment, the significance level, the standard deviation (for quantitative characters) and the type of test. Thus, it is possible to estimate the chance that this is a false-negative result, although it must be admitted that this is rarely done.

In some cases additional information, not contained in the experiment, will be available which may modify these probabilities. With conventional statistical methods, no account is taken of previous knowledge. It is a bit like betting on a horse race where the odds on each horse are exactly the same. However, the chance of winning a bet would be enormously increased if prior information were available on the previous history of each horse. Similarly, with potential carcinogens, there may be good evidence that compounds like compound X are very rarely carcinogenic. Possibly information is available to suggest that among 200 similar chemicals tested, only 20 were found to be carcinogenic. Thus, before starting the experiment the odds are 9 : 1 against it being a carcinogen. In such circumstances, it is sometimes useful to use Bayes theorem (Altman 1991) to combine the information on the prior probability that the compound is a carcinogen, with the results from the experiment. Discussion of this approach is beyond the scope of this book, but there are many statisticians who would strongly urge the more widespread use of Bayesian statistical methods. Unfortunately, these methods are not discussed in many statistical textbooks, and introductory texts on Bayesian statistics are often highly mathematical and not readily accessible to most biologists.

7

Concluding remarks: putting the project together

The aim of this chapter is to suggest ways in which a project can be developed efficiently and economically.

1. Questions, hypotheses and models

Most research projects start with an idea. A new technical advance may make it possible to carry out a research project that previously was not technically feasible, or research will open up a new research area that previously had not been considered to be of particular importance or relevance.

Generating ideas is one of the most creative areas of scientific research, and it requires a good command of the literature, an understanding of technical advances, and a knowledge of progress in disciplines that are peripheral to the area of interest. Cross-fertilization between disciplines often leads to important advances.

Once an idea has been developed and discussed with scientific colleagues, the questions and hypotheses that will need to be addressed can begin to be formulated and the models or techniques that will be used can be considered. This is probably the stage at which the use of *in-vitro* methods or lower organisms rather than animals will need to be considered. If it is decided that the use of animals is unavoidable, then laboratory animal professionals will need to be consulted on choice of species. If mice or rats are to be used, then thought needs to be given to choice of strain, genotype and microbiological quality. If surgical methods are to be used, these will need to be discussed with experts on animal surgery, anaesthesia and analgesia. If the study involves infectious organisms, then suitable isolation facilities will be needed.

2. Ethical acceptability

Preliminary discussions with regulators, such as the Home Office in the UK, should probably take place at this stage. They may be able to indicate broadly whether the project appears to be ethically acceptable, and if not they may suggest alternative approaches that may be more acceptable. Some sort of cost/benefit analysis will need to be considered, where the cost is the pain, suffering, distress or lasting harm done to the animals, and the benefit is the probability that the project will be successful and have potential benefits to humans or other animals if it is successful.

Whether the proposed project should be subjected to a local ethical review (the ERP in the UK, the IACUC in the USA) at this stage, before it is funded, is debatable. On the one hand, the funding organizations may find it more acceptable if it has gone through the review process. On the other hand, if it is not funded then the effort will have been wasted.

3. Designing some preliminary experiments

Controlled experiments are not always needed. If the anticipated effect is large, then small numbers of animals can be used. With a clear-cut result such as death within a short period (though this is to be avoided as an endpoint), there is usually no need for controls as animals do not usually die spontaneously within such a period. Similarly, a transgene may have an effect that is phenotypically obvious, in which case no formal experiment may be necessary to compare it with normal controls, at least with respect to the obvious differences. On the other hand, some effects can be of biological interest but are so subtle that they are difficult to detect even with large, carefully designed experiments. It may be necessary to plan some of the experiments in detail before applying for funding. Funding organizations are increasingly concerned with the ethics of using experimental animals, and will want to ensure that the investigator has given thought to the choice of the most appropriate species and strain and to experimental design and statistics.

If the experimental protocols are complex, it may be a good idea to start with a pilot study using very small numbers of animals in order to ensure that everything works as expected. The results of pilot studies will rarely be publishable, but in many cases they will ensure that animals are not wasted because some part of the experimental protocol was badly designed or inappropriate.

The first full-scale experiment will also need to be planned, preferably in consultation with a statistician. The hypothesis to be tested will need to be stated and treatment groups and dependent variables will need to be specified. The use of factorial designs should be considered. Where there may be a possible choice among different inbred strains, more than one strain could be used in order to identify the most appropriate strain for future studies. The experimental design, such as completely randomized, randomized block or repeated measures will need to be specified. The type of data to be collected will need to be considered, since this and the experimental design will determine the statistical methods to be used. Ideally, some random data of the expected type should be generated and a statistical analysis done to ensure that it is feasible. Things that might go wrong should be considered, including the loss of some animals which may complicate the analysis.

4. Funding

In most cases funding will be sought at about this stage. Success will largely depend on the scientific quality of the idea and the proposed questions, hypotheses and model systems that will be used. However, as noted above, funding organizations may give more weight to applications that have already been through an ethical review process.

5. Starting work

Assuming that the work has been through the ethical review process either before or after obtaining the funding, the real work now starts. Researchers working in the UK will need to ensure that they have the correct Home Office personal and project licences under the Animals (Scientific Procedures) Act 1986. Details are given on the Home Office web site (*www. homeoffice.gov.uk/ccpd/aps.htm*). Other Europeans and Americans will need to consider the legislation in their own country. Obtaining licences may involve attending approved training courses.

Equipment, laboratory space and animal house space will need to be obtained. Reagents and animals need to be ordered. If animals are coming from another laboratory or a commercial breeder, their health status may need to be checked to ensure that they do not bring in any new diseases. The animal house staff will help in making suitable arrangements. Animals may also need to be quarantined. Even when animals are obtained from a reliable source, they will usually need to be acclimatized for one or two weeks to enable them to overcome the stress of settling into new surroundings. Stressed animals may give unreliable results.

Space allocation for animals will need to be at least as good as that given in most regulatory guidelines and codes of practice (Canadian Council on Animal Care 1989, Home Office 1995, Institute of Laboratory Animal Resources 1996). Some thought needs to be given to environmental enrichment. Social species such as mice and rats should normally be housed in groups, though male mice may fight and must be separated. The complexity of the cage environment can often be increased by adding 'furniture'. Various types are commercially available.

Detailed experimental protocols and standard operating procedures (SOPs) should be developed and discussed with the animal house personnel, and reviewed by the statistician. Once everything seems to be in order, the first pilot study can be done. The results can be used to modify the first full experiment, should that be necessary. Finally, the first experiment can be started.

6. Subsequent studies

Each study should be analysed as soon as it is completed so that subsequent experiments can be modified if necessary. Although experimental strategy varies, in many cases an individual experiment is set up to test some relatively simple hypothesis which is clearly stated before the experiment is started. However, once that analysis has been done, it will often be worthwhile to do further exploratory data analysis to see whether the experiment provides any unanticipated results that could provide hypotheses for further investigation. Thus each experiment is used to confirm or refute one or a few hypotheses, and to explore others.

No new principles are involved with subsequent experiments, but in many cases these will not be subject to such critical peer and ethical review, so care needs to be taken to maintain standards. Good working relationships need to be developed with animal house staff and statisticians. New laboratory or animal house staff may need training if the quality of the work is to be maintained.

7. Writing it up

The final step is to write up the results and submit them for publication or to a regulator or grant-giving body. Experiments that are not written up are often a waste of animals, though small pilot studies are usually exempted as their aim is to improve the more definitive experiments. Here, some of the points discussed in Chapter 6 need to be taken into account. Exciting positive findings will obviously be more readily accepted than negative ones. But negative findings are important, and can sometimes be turned into positive statements using power analysis. Referees and editors should be impressed by the use of high quality animals, well designed experiments, good statistical analysis, careful attention to humane details, and claims that are fully supported by the data. Clear text, well laid out tables and good graphs and diagrams should help to ensure acceptance by a high-impact journal.

Appendix 1: Brief notes on statistical packages

Many statistical packages are available. They range in size and complexity from large packages intended to permit virtually any desired statistical calculation—which are usually most suitable for use by people with a good statistical background (for example, SAS or GENSTAT)—to packages designed for the less experienced user (MINITAB) or to provide a defined specialized function (such as StatXact). There are also several spreadsheets and other general software packages that carry out some statistical calculations. A few of the available options, with brief comments, are listed below. However, validation is essential when using any type of software, to ensure that the correct statistical methods are being used in the correct way. It is particularly important if there are missing or unbalanced data. The examples in this book can be used as test data, but are not appropriate for complete validation. All of the major dedicated statistical packages should give reliable results, apart from a few software bugs present in virtually all programs. Each version released usually provides more functionality and a better user interface, so the most recent version should be used wherever possible.

MINITAB

This statistical program was originally designed for teaching statistics, but it has grown into a powerful general-purpose program that is relatively easy to use. It will do most of the calculations and analyses required by working research scientists. Both PC and Macintosh versions are available. Data are entered into data sheets, several of which can be open at the same time, where they can easily be stored and manipulated as necessary, prior to the statistical analysis. The program does all the standard statistical analyses including basic statistics such as means, standard deviations, standard errors, correlation and forming tables of varying complexity with associated chi-squared tests. It does regression analysis, ANOVA and the analysis of covariance, including *post-hoc* comparisons. It will also do sample size estimations using power analysis for comparing means in a one-way ANOVA and for comparing two proportions. It offers a range of non-parametric methods, and provides good presentation-quality graphics that can easily be edited.

There is a cut-down student version which would be suitable for those unable to afford the full version. It would probably be perfectly adequate for many research scientists, though the full version would be better. Further details can be obtained from the MINITAB Web site: *www.minitab.com*. A full trial version can be downloaded from the site and will work for 30 days.

StatXact for Windows

This program does exact non-parametric tests and is particularly useful for sparse categorical data. The chi-squared test, which is often used to analyse tabular categorical data, is only accurate if there are reasonably large numbers, usually more than five, in each group. StatXact uses exact tests such as Fisher's exact test to overcome this problem. With more complex data sets involving tables with several rows and columns, several of which have small numbers, the calculations can be extremely time-consuming. In these situations, StatXact uses Monte Carlo simulation. However, the program also offers a full range of non-parametric tests, including those that are appropriate for continuous data such as the Mann-Whitney and the Friedman tests. The tests and methods used by StatXact are generally not well covered in usual statistical packages such as MINITAB, so this package complements these. StatXact is not likely to be used very often, but when difficult data of this sort are generated, it is almost essential to use a package such as this one to analyse them. Further details are given on the following Web site: *www.cytel.com*.

SAS

SAS is a commercial software package that provides a wide range of validated statistical tools. It is both powerful and versatile, and has especially good data management and manipulation facilities. However, because of the multiplicity of options and facilities and the extensive programming capabilities, it is not simple to use, and the advice and help of a professional statistician familiar with this package may be helpful. Details can be found on the Web site: *www.sas.com/products/sassystem/index.html*.

EXCEL

EXCEL is a spreadsheet that offers some statistical and computing facilities. However, no spreadsheet is capable of doing all of the analyses described in this book. There is an add-in program called StatPlus, which is inexpensive and comes with a textbook (Berk & Carey 1995). It will do one-way ANOVA and two-way ANOVA with equal subclass numbers, as well as a wide range of other statistical techniques, and includes directions for the production and interpretation of residuals plots. However, it has not been used by the authors of this book, and cannot do all the analyses described here.

GENSTAT

GENSTAT is a commercial software package that provides a wide range of validated statistical tools. It is both powerful and versatile, and has especially powerful facilities for analysis of variance. However, because of the versatility of the package, and the need correctly to specify the statistical model, the advice of a professional statistician familiar with this package may be helpful. More details are given on the Web site: *www.nag.com/stats/tt_soft.asp.*

Appendix 2: Further reading

Books on laboratory animal science

Poole T, English P, editors. *The UFAW Handbook on The Care and Management of Laboratory Animals.* 7th edn. Vol. 1: Terrestrial Vertebrates. 864pp. ISBN 0-632-05131-0. Vol. 2: Amphibious and Aquatic Vertebrates and Advanced Invertebrates. 208 pp. ISBN 0-632-05132-9. Oxford, Malden, Paris: Blackwell Scientific, 1999
A large, two-volume comprehensive reference book on all aspects of laboratory animal science. Volume 1 deals with terrestrial vertebrates and Volume 2 with aquatic and amphibious animals. Both have general chapters on environment, animal house design, nutrition and genetics, followed by chapters on individual species.

van Zutphen LFM, Baumans V, Beynen AC, editors. *Principles of Laboratory Animal Science.* Revised edn. Amsterdam, London, New York, Oxford: Elsevier, 2001. 416 pp. ISBN 0-444-50612-8
A handy reference book that covers the principles of laboratory animal science with chapters on legislation, the biology of laboratory animals, standardization, nutrition, genetics, diseases and microbiology and the design of animal experiments. It does not have chapters on individual species.

Wolfensohn S, Lloyd M. *Handbook of Laboratory Animal Management and Welfare.* Oxford: Blackwell Scientific, 1998. 334 pp. ISBN 0-632-05052-7
This is a general introduction to laboratory animal science often used by animal technicians, but it is useful to anyone starting work with laboratory animals. It has chapters on individual species.

Books on experimental design

Clarke GM, Kempson RE. *Introduction to the Design and Analysis of Experiments.* London, Sydney, Auckland: Arnold, 1997. 344 pp. ISBN 0-340-64555-5
A modern book aimed at undergraduates in statistics and mathematics rather than research scientists. As such, it is relatively mathematical.

Cochran WG, Cox GM. *Experimental Designs.* New York, London: John Wiley & Sons, Inc, 1957. 611 pp
A classic book on experimental design, which can be recommended to anyone using more advanced designs. Highly readable with relatively few mathematical formulae.

Cox DR. *Planning Experiments.* New York: John Wiley and Sons, 1958. 208 pp. ISBN 0-471-57429-5
Although published many years ago, this book is still in print and is not in any way dated. It is a readable book aimed directly at the research worker, with few mathematical formulae. However, it also manages to deal with some advanced concepts and experimental designs. Strongly recommended.

Cox DR, Reid N. *The Theory of the Design of Experiments.* Boca Raton, Florida: Chapman and Hall/CRC Press, 2000. ISBN 1-58488-X
An advanced book covering the principles of experimental design, with quite a bit of mathematical notation. However, the text is clear and easy to understand even for the non-mathematician.

Mead R. *The Design of Experiments.* Cambridge, New York: Cambridge University Press, 1988. 620 pp. ISBN 0-521-28762-6
A relatively advanced textbook on experimental design whose author 'has been particularly motivated by teaching postgraduate students specializing in statistics'. It 'requires a sound mathematical background beyond school level'. However, the book emphasizes statistical concepts which should make it possible for the non-mathematical reader to bypass the more complex mathematical details.

Montgomery DC. *Design and Analysis of Experiments.* 4th edn. New York: Wiley, 1997. ISBN 0-471-15746-5
An advanced book covering all aspects of experimental design with an emphasis on industrial experiments, but particularly useful for factorial and fractional designs. Probably more appropriate for a professional statistician than for the average research scientist.

Books on statistics

Altman DG. *Practical Statistics for Medical Research.* London, Glasgow, New York: Chapman and Hall, 1991. 611 pp. ISBN 0-412-27630-5

A textbook aimed at medical researchers with a bias towards work involving humans, but sufficiently general to cover all the biological sciences, including quite advanced statistical concepts. Readable and not too mathematical.

Howell DC. *Fundamental Statistics for the Behavioral Sciences.* Pacific Grove, London, New York: Duxbury Press, 1999. 494 pp. ISBN 0-534-35821-7

A modern textbook that recognizes that virtually all statistical analyses will be done by computer, so emphasizes the importance of understanding the data, choosing appropriate statistical methods, and interpreting the output from statistical packages. However, the book does not specifically cover experimental design.

Maxwell SE, Delaney HD. *Designing Experiments and Analyzing Data.* Belmont, California: Wadsworth Publishing Company; 1989. 902 pp. ISBN 0-534-98233-6

A classic reference book/textbook covering both experimental design and statistics, but with a bias towards the behavioural sciences. Although it has some advanced topics, according to the authors '..the necessary background for the book is minimal'.

Mead R, Curnow RN. *Statistical Methods in Agriculture and Experimental Biology.* London, New York: Chapman and Hall, 1983. 335 pp. ISBN 0-412-24240-0

The aim of this book 'is to describe and explain those statistical ideas which we believe are an essential part of the intellectual equipment of a scientist working in agriculture or on the experimental side of biology'. There is a strong emphasis on experimental design, quite a few formulae, and lots of worked examples.

References

Altman DG (1982) Statistics in medical journals. *Statistics in Medicine* **1**, 59–71

Altman DG (1991) *Practical Statistics for Medical Research.* London, Glasgow, New York: Chapman and Hall

Altman DG, Machin D, Bryant TN, Gardiner MJ, eds. *Statistics with Confidence.* 2nd edn. London: BMJ Press, 2000

Armitage P (1975) *Sequential Medical Trials.* Oxford: Blackwell Scientific Publications

Beck JA, Lloyd S, Hafezparast M, Lennon-Pierce M, Eppig JT, Festing MFW, Fisher EMC (2000) Genealogies of mouse inbred strains. *Nature Genetics* **24**, 23–5

Berk KN, Carey P (1995) *Data Analysis with Microsoft EXCEL.* Pacific Grove, Boston, London: Duxbury Press

Canadian Council on Animal Care (1989) *Guide to the Care and Use of Experimental Animals.* Ottawa, Ontario: Canadian Council on Animal Care

Chvedoff M, Clarke MR, Faccini JM, Irisari F, Monro AM (1980) Effects on mice of numbers of animal per cage: an 18-month study (preliminary results) *Archives of Toxicology* **4**(Suppl.), 435–8

Clarke GM, Kempson RE (1997) *Introduction to the Design and Analysis of Experiments.* London, Sydney, Auckland: Arnold

Cochran WG, Cox GM (1957) *Experimental Designs.* New York, London: John Wiley & Sons, Inc

Cohen J (1988) *Statistical Power Analysis for the Behavioral Sciences*, 2nd edn. Hillsdale, NJ: Lawrence Erlbaum Associates

Cox DR (1958) *Planning Experiments.* New York: John Wiley and Sons

Dixon WJ, Massey FJJ (1983) *Introduction to Statistical Analysis.* Auckland, London, Tokyo: McGraw-Hill International

Everitt BS (1995) The analysis of repeated measures: a practical review with examples. *The Statistician* **44**, 113–35

Festing MFW (1994a) Are animal experiments well designed? In: *Welfare and Science. Proceedings of the 5th Symposium of the Federation of European Laboratory Animal Science Associations, Brighton, 1993* (Bunyon J, ed). London: Royal Society of Medicine Press, 32–6

Festing MFW (1994b) Reduction of animal use: experimental design and quality of experiments. *Laboratory Animals* **28**, 212–21

Festing MFW (1995) Use of a multi-strain assay could improve the NTP carcinogenesis bioassay program. *Environmental Health Perspectives* **103**, 44–52

Festing MFW (1996) Are animal experiments in toxicological research the 'right' size? In: *Statistics in Toxicology* (Morgan BJT, ed). Oxford: Clarendon Press, 3–11

Festing MFW (1997) Fat rats and carcinogen screening. *Nature* **388**, 321–2

Festing MFW (1999) Warning: the use of genetically heterogeneous mice may seriously damage your research. *Neurobiology of Aging* **20**, 237–44

Festing MFW (2000) Common errors in the statistical analysis of experimental data. In: *Progress in the Reduction, Refinement and Replacement of Animal Experimentation*, Vol 1 (Balls M, van Zeller A-M, Halder ME, eds). Amsterdam, New York, Oxford: Elsevier, 753–8

Festing MFW, Fisher EMC (2000) Mighty mice. *Nature* **404**, 815

Finney DJ (1970) In: *Statistics in Endocrinology* (McArthur JW, Colton T, eds). London: MIT Press, 25

Gärtner K (1990) A third component causing random variability beside environment and genotype. A reason for limited success of a 30 year long effort to standardize laboratory animals. *Laboratory Animals* **24**, 71–7

Höglund AU, Renström A (2000) Evaluation of individually ventilated cage systems for laboratory rodents: cage environment and animal health aspects. *Laboratory Animals* **35**, 51–7

Home Office (1995) *Code of Practice for the housing and care of animals in designated breeding and supplying establishments.* London: HMSO

Home Office (2000) *Guidance on the operation of the Animals (Scientific Procedures) Act 1986.* London: The Stationery Office (*www.homeoffice. gov.uk/ccpd/aps.htm*)

Howell DC (1999) *Fundamental Statistics for the Behavioral Sciences.* Pacific Grove, London, New York: Duxbury Press

Institute of Laboratory Animal Resources (1996) *Guide for the Care and Use of Laboratory Animals.* Washington, DC: National Academy Press

Jay GE (1955) Variation in response of various mouse strains to hexobarbitol (Evpal). *Proceeding of the Society of Experimental Biology and Medicine* **90**, 378–80

Kempermann G, Kuhn HG, Gage FH (1997) More hippocampal neurons in adult mice living in an enriched environment. *Nature* **386**, 493–5

Les EP (1972) A disease related to cage population density: tail lesions of C3H/HeJ mice. *Laboratory Animal Science* **22**, 56–60

Lindsey JR, Baker HJ, Overcash RG, Cassell GH, Hunt CE (1971) Murine chronic respiratory disease. *American Journal of Pathology* **64**, 675–716

Lipnick RL, Cotruvo JA, Hill RN, Bruce RD, Stitzel KA, Walker AP, Chu I, Goddard M, Segal L, Springer JA, Myers RC (1995) Comparison of the up-and-down, conventional LD_{50}, and fixed-dose acute toxicity procedures. *Food and Chemical Toxicology* **33**, 223–31

Maclean N (1994) *Animals with Novel Genes.* Cambridge: Cambridge University Press

Malkinson AM (1979) Prevention of butylated hydroxytoluene-induced lung damage in mice by cedar terpene administration. *Toxicology and Applied Pharmacology* **49**, 551–60

Markel P. Shu P, Ebeling C, Carison GA, Nagle DL, Smutko JS, Moore KJ (1997) Theoretical and empirical issues for marker-assisted breeding of congenic mouse strains. *Nature Genetics* **17**, 280–4

Masoro EJ (1993) Nutrition, including diet restriction, in mammals. *Aging Clinical and Experimental Research* **5**, 269–75

Maxwell SE, Delaney HD (1989) *Designing Experiments and Analyzing Data.* Belmont, California: Wadsworth Publishing Company

McCance (1995) Assessment of statistical procedures used in papers in the *Australian Veterinary Journal. Australian Veterinary Journal* **72**, 322–8

Mead R (1988) *The Design of Experiments.* Cambridge, New York: Cambridge University Press

Morton DB, Jennings M, Buckwell A, Ewbank R, Godfrey C, Holgate B, Inglis I, James R, Page C, Sharman I, Verschoyle R, Westall L, Wilson AB (2000) Refining procedures for the administration of substances. *Laboratory Animals* **35**, 1–41

Nadeau JH, Singer JB, Matin A, Lander ES (2000) Analysing complex genetic traits with chromosome substitution strains. *Nature Genetics* **24**, 221–5

Nevison CM, Hurst JL, Barnard CJ (1999) Strain-specific effects of cage enrichment in male laboratory mice (*Mus musculus*). *Animal Welfare* **8**, 361–79

NRC (National Research Council) (1996) *Guide for the Care and Use of Laboratory Animals.* Washington DC: National Academy Press

Papaioannou VE, Festing MFW (1980) Genetic drift in a stock of laboratory mice. *Laboratory Animals* **14**, 11–13

Petrie A, Watson P (1999) *Statistics for Veterinary and Animal Science.* Abingdon, Maiden, Winnipeg: Blackwell Science

Poole T, English P (1999) *The UFAW Handbook on The Care and Management of Laboratory Animals.* Oxford, Malden, Paris: Blackwell Scientific

Porter AMW (1999) Misuse of correlation and regression in three medical journals. *Journal of the Royal Society of Medicine* **92**, 123–8

Russell WMS, Burch RL (1959) *The Principles of Humane Experimental Technique.* Reprinted 1992. Wheathampstead: Universities Federation for Animal Welfare

Schlede E, Mischke U, Roll R, Kayser D (1992) A national validation study of the acute-toxic-class method—an alternative to the LD50 test. *Archives of Toxicology* **66**, 455–70

Silva AJ, Simpson ME, Takahashi JS, Lipp H-P, Nakanishi S, Wehner JM, Giese KP, Tully T, Able T, Chapman PF, Fox K, Grant S, Itohara S, Lathe R, Mayford M, McNamara JO, Morris RJ, Picciotto M, Roder J, Shin H-S, Slesinger PA, Storm DR, Stryker MP, Tonegawa S, Wang Y, Wolfer DP (1997) Mutant mice and neuroscience: recommendations concerning genetic background. *Neuron* **19**, 755–9

Silver LM (1995) *Mouse Genetics.* New York, Oxford: Oxford University Press

Snedecor GW, Cochran WG (1980) *Statistical Methods.* Ames, Iowa: Iowa State University Press

Snell GD, Stimpfling JH (1966) Genetics of tissue transplantation. In: *Biology of the Laboratory Mouse* (Green EL, ed). New York: McGraw-Hill, 457–91

Sprent P (1993) *Applied Nonparametric Statistical Methods.* London, Glasgow, New York: Chapman and Hall

Stokes WS (2000) Reducing unrelieved pain and distress in laboratory animals using humane endpoints. *ILAR Journal* **41**, 59–61

Taylor BA (1996) Recombinant inbred strains. In: *Genetic Variants and Strains of the Laboratory Mouse*, Vol 2 (Lyon MF, Rastan S, Brown SDM, eds). Oxford, New York, Tokyo: Oxford University Press, 1597–1659

Tramontin AD, Brenowitz EA (2000) Seasonal plasticity in the adult brain. *Trends in Neuroscience* **23**, 251–8

van Herck H, Baumans V, Boere HAG, Hesp AMP, van Lith HA (2000) Orbital sinus blood sampling in rats: effects upon selected behavioural variables. *Laboratory Animals* **34**, 10–19

Van Praag H, Christie BR, Sejnowski TJ, Gage FH (1999) Running enhances neurogenesis, learning, and long-term potentiation in mice. *Proceedings of the National Academy of Sciences of the United States of America* **96**, 13427–31

van Zutphen LFM, Baumans V, Beynen AC (2001) *Principles of Laboratory Animal Science*, Revised edn. Amsterdam, London, New York, Oxford: Elsevier

Vesell ES (1968) Factors altering the responsiveness of mice to hexobarbital. *Pharmacology* **1**, 81–97

Waterton JC, Middleton BJ, Pickford R, Allot CP, Checkley D, Keith RA (2000) Reduced animal use in efficacy testing in disease models with use of sequential experimental designs. In: *Progress in the Reduction, Refinement and Replacement of Animal Experimentation*, Vol 1 (Balls M, van Zeller A-M, Halder ME, eds). Amsterdam, New York, Oxford: Elsevier, 737–45

Yates F (1939) The comparative advantages of systematic and randomized arrangements in the design of agricultural and biological experiments. *Biometrika* **30**, 440–66

Index

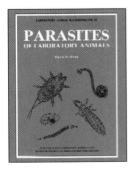